THE ULTIMATE BEGINNER'S GUIDE TO RAISING CHICKENS

BACKYARD CHICKEN KEEPING MADE EASY— RAISE HAPPY HENS AND ENJOY FRESH EGGS EVEN IF YOU'RE A TOTAL BEGINNER!

AVERY SAGE

VIBRANT CIRCLE BOOKS LLC

CONTENTS

INTRODUCTION

Many years ago, I stayed at a bed and breakfast in upstate New York. One sunny morning, I stood in the backyard, a steaming cup of coffee in hand, watching the sunrise cast a golden glow over the owners' small flock of chickens. They clucked contentedly, scratching and pecking at the ground, completely absorbed in their world. It was a simple moment of joy that captured the essence of why so many people are turning to backyard chicken-keeping. The satisfaction of gathering fresh eggs and the connection to nature it brings are unmatched.

This book aims to be your go-to guide for raising chickens, especially if you are just beginning. My goal is to equip you with all the foundational knowledge you need to start and maintain a healthy and productive flock. Whether you're looking to enjoy fresh eggs, practice sustainable living, or simply find joy in caring for these delightful creatures, this book is here to support you every step of the way.

Raising chickens comes with numerous benefits. There's the obvious reward of fresh eggs, but it goes beyond that. By keeping chickens, you take a step towards sustainable living. You create a small ecosystem in your backyard and connect with a more natural way of life. Plus, chickens have a way of bringing unexpected joy. They have personalities, quirks, and routines that can quickly become a beloved part of your daily life.

The structure of this book is designed to guide you through every stage of chicken-keeping. We'll start with selecting the right breed that suits your needs and environment. Then, we'll cover setting up the coop, ensuring it's safe and comfortable. You'll learn how to nurture chicks into healthy adults and manage their health and wellness. Each chapter builds on the last, providing you with a clear roadmap for your journey.

It's natural to have concerns as a beginner. You might worry about failing or not knowing enough. Rest assured, this book provides clear, step-by-step guidance to build your confidence. You'll find practical advice and tips to handle common challenges. Mistakes will happen, but they are part of the learning process. With patience and persistence, your efforts will be rewarded.

This guide focuses on small-scale, backyard operations. It's tailored for those who seek self-sufficiency and sustainable living. The advice is practical and actionable, designed to fit into your life-style and align with your goals.

As you read, I invite you to engage actively with the material. Take notes, reflect on your personal goals, and explore additional resources provided throughout the book. This will deepen your understanding and help you tailor the advice to your unique situation.

In closing, I encourage you to take that first step towards chicken-keeping. Apply what you learn and enjoy the rewarding experience of raising your own flock. Start small, stay curious, and embrace the journey. Before you know it, you'll be gathering fresh eggs and watching your chickens with your own sense of wonder and fulfillment. Welcome to the adventure.

CHAPTER 1
GETTING STARTED WITH BACKYARD CHICKENS

My journey into the world of backyard chickens began at a local farmer's market, witnessing a family's joy as they prepared to welcome their new feathered friends. Their enthusiasm was infectious, highlighting the increasing popularity of raising chickens at home. Beyond the allure of fresh eggs, this endeavor represents a deeper venture into self-reliance, environmental stewardship, and forging a closer bond with the natural world. This chapter delves into the foundational step of this fulfilling hobby: selecting the most suitable chicken breed for your needs.

CHOOSING THE RIGHT CHICKEN BREED FOR BEGINNERS

Selecting the right chicken breed is crucial to your success as a new chicken keeper. It's akin to choosing a pet that fits your lifestyle, yet it's also a commitment to understanding their unique needs and behaviors. Different breeds offer varying benefits, from egg production to temperament and adaptability. For instance, if fresh eggs are your primary goal, consider prolific layers like Rhode Island Reds or Leghorns. These breeds can offer a nearly year-round supply of eggs, providing consistent yields that help reduce trips to the grocery store. Rhode Island Reds are particularly hardy and easy to care for, showing resilience in diverse environments, which makes them ideal for beginners focused on egg production.

When it comes to temperament, having chickens that are friendly and easygoing is imperative, especially if you have children or neighbors close by. Buff Orpingtons and Australorps are known for their calm demeanor, making them excellent ambassadors for the joys of chicken-keeping. Buff Orpingtons, often described as the "golden retrievers" of the chicken world, are gentle, approachable, and dependable in their interactions with both humans and other chickens, making them a delightful addition to any backyard flock.

Climate adaptability is another critical factor. If you live in an area with harsh winters or hot summers, choosing breeds that can withstand temperature extremes will save you headaches down the road. Plymouth Rocks and Wyandottes, for instance, are well-suited for variable climates. Their robust feathering insulates them against the biting cold while still allowing them to remain comfortable and relatively cool during warmer months. These breeds exemplify resilience and adaptability, refining the art of chicken-keeping even in less forgiving climates.

The space you have available and your neighborhood's noise tolerance are also essential considerations. If your backyard is small or if you're concerned about noise and potential nuisances, bantam breeds might perfectly suit your needs. These smaller chickens require less space and tend to be quieter than their larger counterparts, reducing disturbances and allowing easy cohabitation with neighbors. For urban settings where noise can be a considerable issue, quieter breeds like the Australorp or even Silkies—though they require more grooming—can be a suitable choice, bringing charm without excessive noise.

Maintenance is another piece of the puzzle. If you're looking for breeds that thrive with minimal intervention, Sussex and Chantecler may be your best bet. These breeds are considered low-maintenance and adapt well to various environments without needing constant care, making them a favorable choice for beginners. On the other hand, designer breeds like Silkies require regular grooming due to their unique feathering. With their distinctive fluff and affectionate nature, they are often kept as pets rather than production birds but may not be ideal if you're looking for minimal upkeep.

To help you make an informed decision, I've included a breed comparison chart below. This chart visually lays out key traits like egg production, temperament, climate adaptability, space requirements, and maintenance needs. It's designed to guide you in selecting the perfect breed that aligns with your goals and environment.

BREED COMPARISON CHART

Breed	Egg Production	Temperament	Climate Adaptability	Space Requirement	Maintenance Needs
Rhode Island Red	High	Independent	Moderate	Standard	Low
Leghorn	Very High	Active	Moderate	Standard	Low
Buff Orpington	Moderate	Docile	High	Standard	Moderate
Australorp	High	Gentle	High	Standard	Low
Plymouth Rock	Moderate	Friendly	High	Standard	Low
Wyandotte	Moderate	Calm	High	Standard	Low
Bantams	Varies	Quiet	Moderate	Small	Low
Sussex	Moderate	Curious	Moderate	Standard	Low
Chantecler	Moderate	Hardy	High	Standard	Low
Silkies	Low	Friendly	Moderate	Small	High

This chart serves as a quick reference guide to help you compare and contrast potential breeds based on what's most important to you. Whether your focus is on egg production, ease of care, or simply having a group of chickens that get along well with each other and their human caretakers, choosing the right breed will set the stage for a successful chicken-keeping experience.

Imagine selecting your breeds as akin to building a small

symphony of characters in your yard, each contributing their melody to the daily rhythm of life. As the days go by, you'll find your chosen breeds reflecting aspects of your own lifestyle and community, blending into the fabric of your home life with ease.

WHERE TO ACQUIRE YOUR CHICKS

One of the first and most exciting steps is acquiring your first chicks. There are several reliable sources to consider, depending on your goals, budget, and the breeds you're interested in. Each source has its pros and cons, and knowing what to expect can help you make the best decision for your flock.

Local feed stores are a common starting point, especially in the spring. These stores often carry popular breeds of chicks that are suitable for backyard flocks. The benefit of buying from a local store is the ability to see the chicks in person, get advice from staff, and avoid shipping costs or stress on the birds. However, the selection may be limited, and it's important to verify that the chicks are vaccinated and properly cared for.

Another option is ordering chicks from a hatchery. Many hatcheries across the U.S. offer a wide range of breeds and will ship chicks directly to your door. This is a great way to get exactly what you want—whether you're looking for egg layers, meat birds, or rare heritage breeds. Hatcheries often have detailed descriptions and helpful guides, making the selection process easier. Just be aware that chicks are shipped through the postal service and must be picked up quickly to ensure their health and safety.

If you're looking for a more sustainable or community-based option, check local farms, homesteaders, or poultry swaps in your area. Sometimes these sources offer healthy, well-socialized chicks that are already acclimated to your local climate. Plus, connecting with experienced chicken keepers can provide invaluable knowledge and ongoing support as you get started. Whichever route you choose, be sure you're ready with the proper brooder setup and supplies before bringing your chicks home.

UNDERSTANDING LOCAL REGULATIONS AND ZONING LAWS

When I started my chicken-keeping research, I quickly discovered that the most significant hurdle wasn't building the coop or selecting the breeds—it was navigating the maze of local regulations and zoning laws. Just like any other pet, chickens come with their own set of rules and restrictions, often more complex than you might expect. To ensure you're on the right side of the law, it's crucial to research your local ordinances thoroughly. Begin by checking your city or county's official website, where you'll often find a section dedicated to animal and livestock regulations. These sites usually provide detailed information on what's allowed, including the maximum number of hens you can keep and whether roosters are permitted. Roosters are often banned in urban areas due to noise concerns, so if you're dreaming of a picturesque rooster crowing at dawn, you might need to adjust your plans.

In many regions, contacting your local agricultural extension office can be a goldmine of information. These offices are staffed by knowledgeable professionals who can provide insights into both state and local regulations. They often offer resources tailored for backyard poultry enthusiasts, ensuring that your flock complies with all health and safety standards. This step is especially important if you live in a densely populated area, where specific restrictions might apply to protect public health and minimize potential nuisances.

Permitting is another critical aspect to consider. Some areas require you to obtain a permit before you can bring chickens into your yard. The requirements for these permits can vary significantly, so be prepared to provide details about your planned setup, including coop dimensions and location. You may also need to demonstrate that you'll adhere to guidelines regarding waste management and cleanliness, essential for preventing unpleasant odors and potential health hazards.

If you're part of a homeowner association (HOA), you'll need to navigate an additional layer of regulations. HOAs often have their own rules about keeping animals, and chickens may not always be welcome. It's crucial to approach your HOA board with a well-prepared presentation highlighting the benefits of chicken-keeping. Focus on the positive aspects, such as sustainable living practices and the educational opportunities for children. Be prepared to address any concerns they might have about noise, odor, or potential property value impacts. Sometimes, demonstrating a commitment to maintaining a clean and discreet operation can make all the difference.

Once you've figured out all these regulations, maintaining legal compliance becomes your next priority. Keep copies of all relevant documentation organized in a dedicated folder or digital file— permits, correspondence with the city or HOA, and any other paperwork should be easily accessible in case you need to reference them quickly. Establishing good neighbor practices is also vital for staying in everyone's good graces. Let your neighbors know that you're planning to keep chickens and address any concerns upfront. Offering them fresh eggs from time to time can work wonders for fostering goodwill.

Remember that while keeping everything above board might seem daunting initially, it's an essential part of responsible chicken ownership. By diligently researching and understanding local regulations, not only do you protect yourself from fines and legal issues, but you also contribute positively to your community's perception of backyard chicken-keeping. Good legal standing allows you to enjoy your flock without worries, letting you focus on the rewarding aspects of this endeavor: fresh eggs on your breakfast table, the gentle clucking of contented hens, and the satisfaction of knowing you're living just a little bit closer to the earth.

The legal landscape is integral to the harmony within which you and your chickens coexist. By approaching this path with thoroughness and respect for regulations, you are paving the way for a sustainable, enriching relationship between your household and your feathered companions. Allow the process to guide you towards cultivating a neighborhood-friendly routine, one that emphasizes flourishing connections over hassles.

ESSENTIAL EQUIPMENT AND SUPPLIES FOR NEW CHICKEN OWNERS

Walking into the world of chicken-keeping, you'll quickly discover that outfitting your setup with the right gear is crucial. Just like any hobby, having the proper equipment makes all the difference. First and foremost, you'll need reliable chicken feeders and waterers. These are the lifelines for your flock, ensuring they have constant access to nourishment. Choose feeders that minimize spillage and waste, as chickens have a knack for scratching food everywhere. Gravity-fed waterers work well; they provide a steady flow of fresh water while preventing contamination. Investing in durable, easy-to-clean options will save you time and effort in the long run, allowing you to spend more moments enjoying your chickens rather than tending to upkeep.

Equally important are nesting boxes and roosting bars. Nesting boxes offer hens a cozy spot to lay their eggs, safeguarding them from damage and predators. Position these boxes slightly off the ground and fill them with soft bedding like straw or wood shavings to encourage egg-laying. Roosting bars, on the other hand, cater to a chicken's natural instinct to perch while sleeping. Elevate these bars within your coop to keep them safe from ground predators, ensuring a restful roosting experience for your flock. Consider the size of your chickens when installing these essentials; comfort is key to maintaining a happy, productive group of hens.

Now, let's talk about keeping things budget-friendly. If you love a good DIY project, creating your own feeder can be both economical and rewarding. A simple gravity feeder can be constructed using PVC pipes. This design not only saves money but also reduces feed waste by allowing chickens to peck at their leisure without spilling. You can also repurpose household items like old pie tins or shallow pans as temporary feeders or water dishes. Just ensure they are cleaned regularly to prevent contamination.

Safety cannot be overstated when it comes to keeping chickens. Predator-proof locks on the coop door are a must-have, especially if you're in an area with raccoon or fox activity. These cunning creatures can easily open simple latches, so opt for carabiner clips or padlocks to secure entries. Non-toxic cleaning supplies are another vital component of your safety gear arsenal. Chickens are sensitive to chemicals, so using natural cleaners like vinegar or baking soda not only keeps your coop sparkling but also protects your flock's health.

Creating a comprehensive checklist before acquiring your chickens ensures you don't overlook any details. This checklist should cover everything from basic gear to safety items:

CHICKEN-KEEPING ESSENTIALS CHECKLIST

- **Feeders**: Gravity-fed or DIY PVC feeders
- **Waterers**: Gravity waterers or automatic drinkers
- **Nesting Boxes**: One per 3-4 hens, filled with soft bedding
- **Roosting Bars**: Elevated perches for sleeping
- **Predator-Proof Locks**: Secure carabiner clips or padlocks
- **Bedding Material**: Straw, wood shavings, or sand
- **Non-Toxic Cleaning Supplies**: Vinegar, baking soda
- **Insulation Materials**: For cold climates
- **First Aid Kit**: Basic supplies for minor injuries

This checklist acts as a roadmap, guiding you through the initial setup phase and ensuring you're prepared for all aspects of chicken care.

Before you bring home your new flock members, take a moment to reflect on how you'll set up your space. Imagine this: It's a Sunday afternoon, and you're in your backyard setting up the final touches on your coop. You've got a sense of anticipation mixed with excitement because soon you'll be tending to your very own chickens. As you lay down fresh bedding and secure locks on the coop doors, you know these preparations will make all the difference in creating a safe haven for your chickens.

Remember that while you're focusing on practicality, there's room for creativity and personalization, too. Maybe you'll paint the coop in cheerful colors or add whimsical signs that reflect your style. These personal touches enhance not only the appearance but also your connection with this delightful endeavor.

The journey into chicken-keeping is as much about preparation as it is about discovery. Your newfound companions will provide endless opportunities for learning and joy. With the right equipment and a bit of creativity, you're well on your way to creating a haven where chickens thrive and eggs abound.

PLANNING AND DESIGNING YOUR CHICKEN COOP

As you step into the delightful realm of chicken-keeping, an important thing you'll want to consider is how to design a chicken coop that's both functional and inviting. We'll dive into this topic in greater detail in the next chapter, but for now, I want to give you some important points to consider.

Think of the coop as your chickens' home base—a sanctuary where they can eat, sleep, and lay eggs in safety and comfort. At its core, a well-designed coop needs to provide adequate space for each chicken. The general rule of thumb is about 2 to 4 square feet per bird inside the coop, with an additional 8 to 10 square feet in an outdoor run. This ensures they have enough room to move freely without feeling cramped. Adequate space prevents stress, which can lead to pecking and other behavioral issues.

Basic Chicken Coop

Ventilation is another critical element. Good airflow keeps the coop fresh and reduces the risk of respiratory diseases. You can achieve this by installing windows or vents near the roofline, allowing heat and moisture to escape without creating drafts at the birds' level. This setup is akin to the way we open windows in our homes to let in fresh air while maintaining a comfortable temperature. Ventilation not only keeps the air circulating but also helps control humidity levels, particularly in warmer climates, where heat can become oppressive.

When it comes to customizing your coop, the possibilities are endless. Perhaps you have a small yard and are worried about space. Portable coops might be your answer. These mobile units, often called "chicken tractors," allow you to move your chickens around the yard, giving them access to fresh grass while keeping them safe from predators. They're perfect for urban settings where space is at a premium. For those with larger flocks or more land, consider multi-tier designs that maximize vertical space. These designs not only provide ample room for roosting and nesting but also make egg collection more convenient.

Safety and security should remain at the forefront of your planning. Chickens are vulnerable to a host of predators, from raccoons and foxes to hawks and snakes. Reinforced wire mesh barriers are essential for keeping these threats at bay. Always choose hardware cloth instead of chicken wire, as the latter can be easily breached by determined intruders. Ensure all doors and openings have securely latching mechanisms—simple hooks or carabiners work well here —to prevent unauthorized entry by curious critters.

Incorporate user-friendly features to make daily tasks easier, such as external egg collection boxes that save you from entering the coop or feed storage compartments within easy reach. Little conveniences add up, making the routine of caring for your chickens less of a chore and more of an enjoyable part of your daily life. Having a well-thought-out coop design reflects not only efficiency but also the bond you are nurturing with your feathered friends.

To help inspire your coop design, consider practical layout examples that cater to different needs. Visualizing various setups can spark ideas on how best to arrange your coop based on your specific space and flock size. A simple rectangular layout with separate areas for nesting and roosting is a classic choice that suits most backyard flocks. Alternatively, L-shaped or circular designs can offer unique aesthetic appeal while still providing functionality.

Incorporating these elements into your coop design isn't just about meeting the basics; it's about creating an environment where your chickens can thrive. Picture this: It's a crisp morning, and as you sip your coffee, you watch your chickens cluck contentedly within their well-planned home. Their feathers glisten in the morning light as they peck around, exploring their surroundings with curiosity. The satisfaction of knowing you've provided a safe haven where they can express natural behaviors without fear is immensely rewarding.

As you embark on this planning phase, allow yourself the freedom to be creative yet practical. Your coop should reflect not only the needs of your chickens but also your personal style and the constraints of your environment. Personal touches like adding a small garden bed nearby or painting the coop in vibrant colors can transform a simple structure into a delightful backyard feature.

Remember that each decision you make now lays the groundwork for happy and healthy chickens. By ensuring ample space, effective ventilation, thoughtful customization, and robust security measures, you're setting the stage for success in your chicken-keeping endeavor. The journey of designing and building a coop may seem daunting at first, but with each step, you'll find yourself more connected to this fulfilling lifestyle choice—one that's not just about sustenance but about nurturing life and appreciating simplicity.

SETTING UP A SAFE AND COMFORTABLE BROODER

Setting up a brooder for your chicks is like crafting their first cozy home. It's where they'll eat, sleep, and grow until they're ready to join the others in the coop. The brooder is essentially a warm, secure space that mimics a mother hen's care. Start with the heat source, which is vital for young chicks who can't regulate their body temperature. Heat lamps are a popular choice due to their efficiency and reliability. Hang the lamp securely above the brooder, ensuring it's safely out of reach from flammable materials. Alternatively, consider using a radiant heat plate—a safer option that replicates the warmth of a mother hen by allowing chicks to huddle beneath it. Both methods will keep your chicks snug and comfortable.

Cutaway Image of a Brooder

Now, let's talk bedding. Bedding absorbs waste, reduces odor, and offers a soft surface for the chicks to walk on. Pine shavings are an excellent choice—affordable, absorbent, and easy to clean. Avoid cedar shavings; they can be toxic to chicks. Some people use straw or shredded paper, but these can be less absorbent and might require more frequent changes. Layer the bedding about two inches thick over the floor of the brooder to ensure adequate cushioning.

Maintaining the right temperature and humidity in the brooder is crucial for chick health. During their first week, chicks need a cozy environment of about 95°F. Reduce this temperature by 5°F each week until they're ready for the coop. A simple thermometer placed at chick level will help monitor this precisely. Pay attention to humidity, too—keeping it around 50% is ideal. Use a hygrometer to track humidity levels, adding a shallow dish of water to increase moisture if needed.

Brooder maintenance involves regular cleaning to prevent disease and promote healthy growth. Chicks are messy little creatures, so establish a cleaning schedule early on. Remove droppings daily and change bedding at least once a week. Keeping surfaces clean is essential for preventing respiratory issues. Disinfect the brooder every few weeks using a solution of water and vinegar or mild soap—avoid harsh chemicals that could harm your chicks.

Monitoring chick behavior gives you invaluable insight into their well-being. Chicks naturally chirp and explore, but excessive peeping might indicate discomfort. If you notice them huddling under the heat source, they're likely cold. Conversely, if they're scattered far from it, they might be too warm. Adjust the heat source accordingly to maintain their comfort zone. Regularly observe their activity and feed consumption; these are indicators of health and contentment.

Keeping an eye out for signs of illness is also crucial. Healthy chicks are active with bright eyes and clean, fluffy feathers. Look for symptoms such as lethargy, labored breathing, or drooping wings—these may suggest health issues requiring prompt attention. If you spot any signs of illness, separate the affected chick immediately to prevent spreading disease and consult with a vet experienced in poultry care.

Establishing a nurturing environment within the brooder is essential for laying the groundwork to raise strong, healthy chickens. As each day passes, you'll observe not just physical growth but also the development of a deeper bond between you and your chicks. This formative stage is critical, as it's during this time that these tiny, delicate creatures gradually evolve into hearty, robust birds, fully equipped and prepared for the broader world beyond the confines of their first home. Watching this transformation unfold is a remarkable journey, filled with moments of learning, care, and connection that underscore the importance of your role in their early life. This nurturing process ensures that by the time they are ready to transition to the coop, they are not only physically prepared but also have developed a trust in their environment and in you, their caretaker, setting the stage for a successful integration into the flock.

In crafting your brooder setup, remember that flexibility is key. Each flock has its quirks and preferences, so adjusting based on their behavior will ensure you meet their needs effectively. It's all about finding that sweet spot where science meets instinct— creating a space that's both practical and comforting for your chicks as they take their first steps in life.

INTRODUCING YOUR FIRST FLOCK TO THE BACKYARD

Bringing your chickens into their new outdoor environment is a moment of excitement and anticipation. It's akin to watching children explore a playground for the first time. But just like kids, chickens need time to acclimate to their surroundings gradually. When you first introduce them to the backyard, short, supervised visits are crucial. Start by allowing them to explore a small, contained area where they can get used to the sights, sounds, and smells of their new world. These initial outings should be brief, perhaps just ten to fifteen minutes, slowly increasing as they become more comfortable and confident.

As your chickens grow accustomed to their new environment, you can gradually expand their roaming area. This gradual expansion helps prevent overwhelming them and allows them to establish a sense of security. It's akin to giving them training wheels before they're ready to ride freely. You'll notice that, over time, they'll begin to venture further and further, pecking and scratching with increasing curiosity. This process not only aids in their acclimation but also helps them develop a healthy routine of foraging and exploring.

Flock dynamics play a significant role in how your chickens interact with one another. Chickens naturally establish a pecking order—a social hierarchy that determines dominance within the group. It's fascinating to observe these dynamics unfold as they assert their roles through subtle behaviors. You might see one chicken strutting confidently while others defer by stepping aside. This is normal behavior and usually resolves itself without intervention. However, if aggression becomes an issue, there are strategies to minimize it, and we will address this topic thoroughly later in the book.

Predator awareness is an essential aspect of keeping your flock safe. Every backyard has its own set of potential threats, ranging from neighborhood cats and dogs to more elusive foes like raccoons and hawks. Recognizing these local predators is the first step in protecting your flock. Pay attention to what's prevalent in your area —ask neighbors or check local wildlife resources for common threats. Implement deterrent measures, such as installing motion-activated lights or decoy owls to discourage predators from approaching. Secure fencing and covered runs provide additional layers of protection.

Sharing stories from fellow chicken keepers can offer both motivation and reassurance. Take, for instance, the experience of a friend who initially struggled with integrating a new flock into her busy urban backyard. She started with short outings, just as we've discussed, but noticed her flock was hesitant to explore beyond their coop. With patience and a few enticing treats scattered around their enclosed area, she gradually encouraged them to roam further each day. Within a couple of weeks, her chickens were confidently strutting across the yard, engaging in natural behaviors like dust bathing and foraging.

Another fellow chicken keeper faced challenges with predator threats in his rural setting. He discovered raccoons frequenting his property at night. By using solar-powered predator lights that mimic eyes glowing in the dark, he effectively deterred these nocturnal intruders without harming them or his chickens. These real-life accounts illustrate that while challenges may arise, there are always creative solutions to navigate them successfully.

Remember that each chicken keeper's experience is unique, shaped by individual circumstances and environments. But the underlying principles of gradual acclimation, understanding flock dynamics, and ensuring predator protection are universal keys to success. As you embark on this adventure with your flock, embrace the learning curve with openness and curiosity. You'll find that watching your chickens thrive in their new home brings a sense of accomplishment and joy that few other hobbies can match.

As you conclude this foundational chapter in your chicken-keeping adventure, reflect on the steps you've taken so far and the knowledge you've gained. The initial stages of introducing your flock are filled with discovery—not just about chickens but also about yourself as a caretaker. Each moment spent observing their antics and nurturing their growth reinforces the bond between you and these feathered companions. It's a rewarding experience that promises fresh eggs along with endless moments of amusement. With patience, care, and an open heart, you are well on your way to becoming a confident and capable chicken keeper.

CHAPTER 2

BUILDING THE PERFECT COOP

On a breezy afternoon, I found myself researching coop designs, my mind brimming with ideas. It was then I realized that creating a safe haven for my flock was more than just a construction project—it was a labor of love. You see, the coop isn't merely a shelter; it's a fortress protecting your chickens from the sly antics of neighborhood predators. Crafting a predator-proof coop demands attention to detail and an understanding of potential threats.

UNDERSTANDING THE BASICS

A well-designed chicken coop is essential for keeping your flock safe, healthy, and productive. At the heart of the setup is the coop itself, a secure, weatherproof shelter where chickens can sleep, lay eggs, and take refuge from predators and harsh weather. The coop should be well-ventilated but draft-free, and it must provide enough space for each bird—generally about 2–4 square feet per chicken inside the coop. A solid floor, tight-fitting doors, and hardware cloth over any openings help keep out predators like raccoons, snakes, and weasels.

Connected to the coop is the chicken run, an outdoor area where chickens can roam, scratch, peck, and enjoy the sunlight. This space should be enclosed with strong wire or mesh to prevent escapes and deter predators, particularly from above.

Inside the coop, nesting boxes are essential for egg-laying hens; these should be filled with clean bedding and placed in a quiet, dim area of the coop to encourage hens to lay there consistently. Each box can serve about 3–4 hens.

Finally, roosting bars are critical for nighttime, as chickens prefer to sleep off the ground. These bars should be placed higher than the nesting boxes to prevent chickens from sleeping where they lay eggs, and they should be wide enough for the birds to comfortably grip while they sleep. Together, these components create a safe and functional environment for your backyard flock.

Interior of a Coop, with Nesting Boxes and Roosting Bars

BUILDING OPTIONS

Many readers will be highly experienced in various types of DIY projects, and may be eager to take on the challenge of designing and building a chicken coop from scratch. I will share tips for the DIYers later in this chapter. Others might find the prospect of doing this on their own from the ground up more daunting. Don't worry, I've got you covered.

Here are two excellent online resources that provide detailed plans and instructions for building a chicken coop, ranging from beginner-friendly to more advanced builds:

1. Backyard Chickens – Coop Plans & Designs

- **URL:** https://www.backyardchickens.com/articles/category/chicken-coops.12/
- **What it offers:** Hundreds of user-submitted chicken coop plans with photos, materials lists, and step-by-step instructions. Great for all skill levels.
- **Why it's useful:** Real-life examples from chicken keepers, with feedback and variations.

2. The Garden Coop

- **URL:** https://www.thegardencoop.com
- **What it offers:** Professionally designed chicken coop and run plans available for purchase, including the popular "Garden Coop" and "Garden Ark."
- **Why it's useful:** Well-designed plans with precise instructions, diagrams, and tips—ideal for secure, predator-proof coops.

And if you tend to shy away from DIY building projects altogether, fear not, there are sources for prefabricated options as well. Here are two of the best.

1. Omlet

- **Website:** https://www.omlet.us
- **Direct link:** https://www.omlet.us/chicken-coops/
- **What they offer:** Modern, weatherproof, and predator-resistant chicken coops like the *Eglu Cube* and *Eglu Go*. Modular designs with options for wheels, runs, and accessories.
- **Why it's great:** Known for durability, easy cleaning, and minimal maintenance. Great for urban or suburban backyards.

2. Tractor Supply Co.

- **Website:** https://www.tractorsupply.com
- **Direct link:** https://www.tractorsupply.com/tsc/catalog/coops-pens
- **What they offer:** A wide selection of prefab wooden chicken coops, ranging from small backyard setups to larger walk-in coops.
- **Why it's great:** Offers in-store pickup or delivery. Good variety in budget, style, and size.

STEPS TO BUILDING A PREDATOR-PROOF COOP

First and foremost, let's delve into the materials you'll need to make your coop as secure as possible. Hardware cloth is your best friend here. Unlike chicken wire, which can be easily breached by determined predators, hardware cloth offers robust protection. Its small mesh size prevents sneaky paws from reaching in and causing harm. Make sure to cover all windows and vents with hardware cloth, securing it tightly to avoid any gaps that crafty critters might exploit. As an added layer of security, consider using heavy-duty staples or screws with washers rather than nails, which can be pried loose over time.

Next on the list, consider installing underground barriers to deter digging animals, like raccoons and foxes. Burying hardware cloth or concrete blocks about a foot deep around the perimeter of your coop creates an impenetrable barrier. This extra step might seem tedious, but it's worth every bit of effort when you see your chickens safe and sound. To further enhance this, some chicken keepers have found success by including sharp-edged materials or coarse gravel beneath the barrier, adding a dissuasive texture that makes digging uncomfortable for potential predators.

When it comes to construction materials, pressure-treated wood is a solid choice for durability. It withstands the elements and stands firm against the test of time. However, it's crucial to avoid using wood treated with chemicals harmful to chickens if they decide to peck it. Non-toxic finishes are available and should be used to ensure their safety. Ensure all doors are equipped with reinforced locks and latches. Opt for two-step latches or carabiners that raccoons can't figure out—because they will try! A determined predator can open simple locks, so this added security measure is crucial.

Now, let's delve into some design strategies that naturally deter unwanted visitors. Elevating your coop is an effective way to keep ground predators at bay. By raising the structure off the ground, you eliminate easy access points for digging animals, providing an additional layer of security. Moreover, incorporating a slanted roof not only aids in water runoff but also helps prevent climbing creatures from gaining a foothold. This simple architectural feature can significantly reduce the risk of predators scaling your coop. For a more efficient design, solar or motion-activated deterrent lights can be installed at vantage points to frighten nocturnal invaders.

PRACTICAL EXAMPLES AND REAL-WORLD TECHNIQUES

Real-world experiences often provide invaluable lessons. I recall a fellow chicken keeper who initially underestimated the persistence of local raccoons. After several frustrating attempts to secure her coop with basic locks and chicken wire, she switched to hardware cloth and installed underground barriers. The transformation was remarkable—no more missing chickens or midnight disturbances. Her story is a testament to the effectiveness of these preventive measures. To enhance her setup, she later added perimeter alarms that alert her when disturbances occur, offering real-time protection.

Whether or not you live in a predator-heavy area, investing time in these protective strategies ensures long-term peace of mind. A friend I know, who also faced numerous predator challenges, added additional layers of security by installing motion-activated floodlights around the coop's perimeter. These lights, coupled with his existing reinforcements, became a powerful deterrent. Knowing that predators would flee the startling sudden brightness provided him peace and his chickens tranquility. As dusk settles, these lights activate in gentle increments to avoid startling the chickens but effectively ward off unwanted visitors.

CASE STUDY: LESSONS LEARNED FROM PREDATOR-PROOFING SUCCESSES

Imagine your coop as Fort Knox for chickens. Another friend of mine, living in a rural area with frequent predator sightings, faced numerous challenges before achieving success. Initially, he also used chicken wire and basic latches, only to discover raccoons easily breached them. After switching to hardware cloth and carabiner clips, his flock remained undisturbed. The small investment in these security measures paid off in peace of mind and happy chickens. Furthermore, he introduced metal flashing around the base of the coop, preventing rats from gnawing their way inside. It was a revelation, showing just how intricate a predator's attempts can be and how prepared one must be to outsmart them. His journey underscored the principle that in predator-proofing, redundancy isn't just a precaution—it's a necessity.

An additional consideration he implemented was a perimeter noise deterrent system that emitted high-pitched frequencies only audible to animals. This innovative approach drastically reduced the number of predator visits, adding another successful layer to his protection strategy.

By following these guidelines and learning from real-life examples, you can build a coop that not only meets your needs but also stands strong against potential threats. Your chickens will show their appreciation with clucks of contentment as they thrive in their safe haven. Remember that each step you take in fortifying your coop serves as an investment in their well-being, allowing you to enjoy the many rewards of backyard chicken-keeping without worry.

OPTIMAL COOP VENTILATION AND TEMPERATURE CONTROL

Imagine this: it's a hot summer day, and your coop feels like an oven. The air is heavy, and your chickens are panting, desperately seeking relief. Proper ventilation is crucial in preventing such scenarios. It ensures your flock remains healthy by allowing fresh air to circulate, expelling moisture, and keeping harmful bacteria at bay. Place windows or vents high up to let warm air escape without creating drafts at the birds' level. In warmer regions, consider installing fans or wind-driven turbines to enhance airflow. These simple additions can make a world of difference in keeping your coop fresh and your chickens comfortable. To monitor the efficiency of your ventilation system, integrate humidity and temperature sensors that offer real-time data to adjust ventilation accordingly.

Temperature control is another key factor in maintaining a healthy coop environment. Insulated walls can be a game-changer, especially in regions with extreme weather conditions. They keep the coop cool in summer and warm in winter. For those chilly months, a heat lamp can offer much-needed warmth, but safety is paramount. Ensure it's securely placed, away from flammable materials, to prevent accidents. Consider using a ceramic heat emitter instead of traditional bulbs. They provide consistent heat without the risk of shattering. Some keepers vouch for heated flooring options, which evenly distribute warmth and alleviate the need for overhead heat sources, providing another point of comfort for your chickens.

STRATEGIES TO MANAGE MOISTURE AND AIR QUALITY

Managing moisture levels is essential to prevent respiratory issues in chickens. High humidity can lead to damp bedding and mold growth, creating an unhealthy environment. Drip edges and gutters on the roof can effectively direct rainwater away from the coop, minimizing moisture accumulation. Inside, choose bedding materials like straw or wood shavings that absorb moisture well and can be changed easily. Regularly check for wet spots and replace damp bedding promptly to maintain a dry and comfortable living area for your flock. Implement moisture-absorbing pouches or dehumidifiers in especially rainy climates to combat high humidity effectively.

Insights from seasoned chicken keepers and veterinarians underscore the importance of environmental control. One experienced keeper shared how she battled respiratory issues in her flock by tweaking ventilation and moisture management strategies. By adding extra vents and switching to sand bedding, she noticed a significant improvement in her chickens' health. Veterinarians also emphasize regular air quality checks, recommending that coop owners pay attention to ammonia levels—a common byproduct of chicken droppings that can irritate respiratory systems if allowed to build up. Installing ammonia monitors regularly within the coop can alert you to unhealthy levels, allowing for prompt action to ensure the well-being of your flock.

Incorporating these expert insights into your coop setup can significantly impact your chickens' well-being. It's not just about following guidelines; it's about observing your flock and making necessary adjustments to ensure they thrive. For instance, if you notice condensation on the windows or a musty smell, these are signs of inadequate ventilation or high humidity levels that need immediate attention.

A well-ventilated coop with stable temperature control will not only enhance your chickens' comfort but also boost egg production and overall flock health. The effort you put into creating an optimal environment reaps rewards in the form of happy, productive birds. As you fine-tune your coop's setup, remember that small changes can lead to significant improvements. Each adjustment you make brings you closer to providing the best possible home for your feathered friends.

Your chickens rely on you to create a space where they can thrive, lay eggs consistently, and live comfortably. By focusing on ventilation, temperature regulation, and moisture control, you're setting the stage for success in your chicken-keeping endeavors. Ultimately, it's about creating a balanced environment that caters to your flock's needs while ensuring their safety and happiness.

CREATING A SECURE OUTDOOR RUN FOR YOUR CHICKENS

Imagine your chickens enjoying the fresh air and sunshine in their secure outdoor run, happily pecking and scratching away. It's crucial to provide them with a safe space to roam, where they can indulge in their natural behaviors while remaining protected from potential threats. When constructing your chicken run, one of the first decisions you'll face is choosing the right fencing material. Welded wire offers a sturdy barrier that keeps most ground predators at bay. Its rigid structure and small mesh size prevent unwanted guests from squeezing through gaps, ensuring your flock's safety. On the other hand, electric fencing provides an extra layer of security, delivering a mild shock to deter curious critters. It's an effective option for those living in areas with persistent predator problems.

DESIGNING ACCESS POINTS AND ENHANCING SECURITY

Designing access points is another key consideration when setting up your run. You'll want to create entrances that are easy for you to use but challenging for predators to breach. Consider installing a simple latch system that allows you to open the gate with one hand while carrying feed or water with the other. Make sure the gates are secure and fit snugly within the frame to prevent any sly creatures from slipping through. A double-door entryway can add an extra level of security, giving you peace of mind knowing your chickens are safe. Implement technology such as automated locks that respond to your presence or specific times, adding a high-tech touch to your chicken care practices.

While solid fencing forms the backbone of your run's defenses, additional deterrents can further increase its security. Motion-activated lights or sprinklers serve as excellent surprise tactics against nocturnal predators. A sudden flash of light or burst of water can startle and discourage them from approaching your chickens' sanctuary. Overhead netting is crucial for protecting against aerial predators like hawks and owls. By covering the top of your run with durable netting, you prevent these birds of prey from swooping down and making off with one of your hens.

Beyond safety, it's important to enrich your chickens' environment to stimulate their natural behaviors. Adding areas for dust bathing is a must; chickens love nothing more than rolling around in loose sand or peat, cleaning their feathers and keeping parasites at bay. These little dust baths can be as simple as a shallow container filled with sand or an area of the run dedicated to this purpose. Providing perches and climbing structures encourages exercise and satisfies their instinctual need to roost above ground level. Repurposing branches or constructing wooden ladders gives them something to climb on, promoting physical activity and mental stimulation.

To help visualize your setup, it might be helpful to consider different layout options for efficiency and security. Picture a rectangular run with a central access path, allowing easy reach to all areas for cleaning and maintenance. The perimeter could feature reinforced fencing paired with strategically placed deterrents like lights or sprinklers at each corner. Inside this safe zone, designated areas for dust bathing alongside scattered perches create an enriching environment that caters to your chickens' needs. Enhance the landscape with varying levels, such as gentle inclines or raised platforms, providing your chickens with stimulating topography similar to their natural habitat.

As you plan your outdoor run, remember that each choice you make contributes to both the safety and happiness of your flock. No detail is too small when it comes to ensuring their well-being; even the material you choose for the ground can have significant impacts. Consider a mix of gravel and sand for the flooring of your run. Gravel promotes drainage, which helps keep the area dry after rain, while the sand provides a comfortable substrate for dust bathing. Balancing these elements ensures not only their well-being but also enhances your experience as a chicken keeper. Watching your chickens thrive in a secure environment brings immense satisfaction and joy, knowing you've created a space where they can live freely yet safely.

Establishing a secure and enriching outdoor run allows your chickens to explore their surroundings confidently while remaining protected from potential hazards. This harmonious blend of security measures and environmental enrichment fosters a thriving backyard flock that rewards you with contentment and productivity.

DIY COOP DESIGNS AND CREATIVE SOLUTIONS

Building a chicken coop offers a wonderful opportunity to let your creativity shine. It's a chance to tailor a space that meets both the needs of your chickens and your personal style. One way to make your coop stand out is by using recycled materials. Not only does this approach save money, but it also promotes sustainability. Old wooden pallets, for instance, can be deconstructed and repurposed into sturdy walls or flooring. They provide a rustic charm while being budget-friendly. With a little imagination, you can transform these simple materials into a charming coop that blends seamlessly with your backyard.

PERSONALIZATION AND COST-EFFECTIVE DESIGN

Personalization comes into play when you add decorative touches like paint or murals. Imagine bright colors or whimsical designs adorning your coop, turning it into a backyard centerpiece. A splash of color can bring joy to your outdoor space, and painting the coop with themes or patterns makes it uniquely yours. Some folks even involve their kids in the painting process, creating a fun family project that everyone can enjoy. It's amazing how a few brushstrokes can breathe life into a simple structure, making it not just a coop but a piece of art.

For those on a tight budget, inventive solutions abound. Pallet wood isn't just for walls; it can be used to construct the entire framework of your coop. This method is cost-effective and still allows for plenty of customization. Repurposing old furniture as nesting boxes is another clever idea. An unused dresser, for example, can be converted into cozy nesting spaces by removing the drawers and adding bedding. These creative hacks ensure functionality without breaking the bank, demonstrating that resourcefulness is key in building an affordable yet efficient coop.

Space constraints? No problem! Modular coops offer an ideal solution. These designs allow you to expand as needed, adding sections or levels as your flock grows. It's like building blocks for adults, giving you the flexibility to adapt your coop over time. Integrating your coop with existing structures, like garden sheds, is another smart move for maximizing space. By sharing walls or roofs, you reduce material costs and create a seamless connection between different areas of your yard.

Let's not forget the brilliant ideas submitted by fellow chicken keepers who have mastered the DIY approach. One reader shared their success with a vertical coop design that utilizes minimal ground space while providing plenty of room for chickens to roost and nest above eye level. This design is perfect for urban settings where land is limited. Another reader crafted a charming coop using reclaimed barn wood, creating a rustic yet functional space that blends beautifully with its surroundings. These examples show that with a little ingenuity and effort, you can create something truly special.

Incorporating natural elements into your design can also enhance the coop's aesthetic appeal and functionality. Use branches or logs for perches inside the coop, adding texture and variety to the environment. These natural materials are not only cost-effective but also encourage chickens to exercise their natural behaviors by climbing and perching at different heights. Introduce small plants or herbs around the coop's exterior; certain herbs, like lavender or mint, help deter pests and add a delightful fragrance to the air. Creating a DIY chicken coop is about more than just providing shelter; it's about creating a space that reflects your personality and meets the needs of your chickens. Whether you're repurposing materials or adding artistic touches, each decision contributes to a unique and inviting environment. The process of building your coop becomes an adventure in itself, filled with opportunities for innovation and personal expression.

As you embark on this creative endeavor, remember that each coop is as unique as its owner. Your choices in materials, design, and decoration breathe life into the structure, transforming it from mere shelter into a cherished part of your home landscape. With thoughtful planning and a touch of imagination, your DIY coop will stand as a testament to both your ingenuity and your commitment to sustainable living.

Your chickens will appreciate the effort you've put into creating their home, responding with contented clucks and fresh eggs as thanks for providing them with such a delightful abode. This balance of practicality and creativity ensures that both you and your flock thrive, enjoying the fruits of your labor in a space that's entirely your own.

ESSENTIAL COOP MAINTENANCE TIPS FOR BEGINNERS

Waking up to the soft clucks of your chickens and the promise of fresh eggs is a wonderful way to start the day. But to keep your coop running smoothly, a little maintenance goes a long way. Let's break down a simple routine that keeps your coop clean and healthy without overwhelming you. On a daily basis, remove droppings from under the roosting bars. This quick task helps reduce odor and keeps pests at bay. Freshen up the water supply, and make sure your chickens have plenty of clean water to drink. A simple rinse of the waterer can prevent any buildup of bacteria, which is crucial for maintaining flock health. These small daily habits ensure your flock stays happy and your coop remains pleasant.

Weekly, take some time to inspect your coop for any signs of damage. Look for loose boards, gaps in walls, or anything that might compromise its security. Pests can find tiny openings to sneak through, so sealing these up immediately prevents bigger problems down the road. Weekly checks also involve looking for pest infestations. Mites and lice are tiny but can cause big problems if left unchecked. Observe your chickens for signs of over-preening or feather loss, as these might indicate an unwelcome invasion. A proactive approach keeps your chickens comfortable and reduces stress, leading to better egg production.

When it comes to effective cleaning methods, natural options like vinegar or baking soda are your best bet. Vinegar is excellent for disinfecting surfaces without exposing your chickens to harsh chemicals. Mix it with water for a simple spray that can be used on walls and nesting boxes. For deeper cleaning sessions, pressure washing can be immensely effective. It blasts away grime and stubborn dirt from the coop's exterior, leaving it looking fresh and new. Just be sure to remove your chickens first and allow the area to dry completely before letting them back in. This method saves time and effort, making thorough cleanings less of a chore.

PREVENTIVE AND SEASONAL MAINTENANCE

Keeping your coop in tip-top shape involves knowing how to spot signs of wear and tear before they become major issues. Regularly check hinges, latches, and locks to ensure they work smoothly. Seasonal maintenance is equally important, particularly as the weather changes. In the fall, prepare for winter by checking insulation and sealing any drafts that could let in cold air. Weather strips can be an inexpensive and highly effective way to further seal these gaps. In spring, reinforce areas that might have weakened during winter storms or heavy rains. Paying attention to these details fortifies your coop against the elements, preserving its structure and protecting your flock.

Troubleshooting common maintenance issues doesn't have to be daunting. Leaks are often caused by aging roof materials or clogged gutters. If you spot water inside the coop after a rainstorm, inspect the roofline for gaps or cracks and patch them up promptly. Drafts can be tricky; they often sneak in through unnoticed crevices. Feel around on a windy day to locate drafts and seal them with weatherproof tape or caulking. Handling infestations requires diligence and quick action. If mites or lice are discovered, a thorough cleaning followed by dusting your chickens with diatomaceous earth can help eradicate these pests naturally.

A checklist can be a lifesaver when it comes to keeping track of maintenance tasks. Consider drafting one that outlines daily, weekly, and monthly responsibilities, so nothing falls through the cracks. This checklist guides you through routine upkeep while acting as a reminder for less frequent tasks like deep cleaning or structural inspections.

MAINTENANCE CHECKLIST

- **Daily:** Remove droppings, refresh water supply
- **Weekly:** Inspect for damage, check for pests
- **Monthly:** Deep clean with vinegar solution or pressure wash
- **Seasonal:** Weather-proofing adjustments (seal drafts in fall, reinforce in spring)

Embracing routine maintenance ensures your coop remains a safe haven for your chickens. It might seem like a lot at first glance, but once these tasks become part of your regular schedule, you'll find they blend seamlessly into your day-to-day activities. The satisfaction of knowing you've created and maintained a clean, secure environment for your flock is incredibly rewarding. Whether you're scrubbing down walls or tightening loose screws, each effort contributes directly to the well-being of your chickens—and that makes every bit of work worthwhile.

INCORPORATING NATURAL LIGHTING AND SHADE

Over time, I've come to realize the immense benefits of natural lighting for our feathered friends. Natural light plays a pivotal role in stimulating egg production and promoting overall chicken health. Positioning windows strategically in your coop can harness this light effectively. If you live in the Northern Hemisphere like me, aim to install windows on the south-facing side, where they capture the most sunlight throughout the day. This positioning ensures your chickens enjoy the benefits of natural light without overheating.

IMPLEMENTING LIGHT AND SHADE SOLUTIONS

Installing skylights or clear roof panels can further enhance this effect, bathing your coop in gentle, diffused sunlight that mimics their natural environment. Chickens are incredibly perceptive to light, and these features can help regulate their laying cycles and mood. It's remarkable how a few well-placed windows or panels can transform the coop into a sunlit haven, encouraging your hens to lay consistently and remain active and healthy.

However, too much sun exposure can lead to heat stress, especially during those blazing summer months. Providing ample shade is crucial for keeping your flock comfortable. Planting trees or shrubs around the coop offers natural shade that cools the area while adding beauty to your backyard. The foliage not only provides shelter from the sun but also creates a natural habitat for insects, adding a touch of biodiversity that your chickens will love exploring.

When immediate shade is needed, consider using shade cloths or tarps for temporary coverage. These versatile solutions can be draped over the coop and run, offering respite from the sun's intensity. They are easy to adjust and can be removed when the weather cools down. Combining permanent and temporary shading solutions ensures your chickens have access to cooler areas throughout the day, reducing their stress levels and promoting wellbeing.

In the evenings or during darker days, energy-efficient lighting alternatives come into play. Solar-powered lights offer an eco-friendly option for illuminating coops without additional electricity costs. Placing them strategically around your coop can provide just enough light for evening checks or late-night egg collection. LED bulbs are another sustainable choice, offering bright illumination with minimal energy consumption. These modern options help maintain a gentle balance between natural and artificial lighting, ensuring your flock enjoys a consistent environment. Automated timers can regulate artificial lighting, simulating the natural sunrise and sunset to minimize disruption to your flock's circadian rhythm.

Taking inspiration from fellow chicken keepers can provide valuable insights into effective lighting and shading solutions. A neighbor of mine cleverly utilized old window frames fitted with clear polycarbonate sheets as skylights in her coop. The result was a bright, airy space that stayed cool, even during peak summer heat. Another friend planted fast-growing shrubs around her run, creating dappled shade that transformed her chicken yard into a lush oasis. These real-world examples demonstrate that with creativity and a bit of effort, you can achieve an ideal balance of light and shade in your setup.

By considering both natural light and shading techniques, you create an environment where your chickens thrive. They enjoy the benefits of sunlight for health and egg production while having cool retreats to escape the heat. Balancing these elements is key to fostering a happy and productive flock.

As we wrap up this chapter on building the perfect coop, remember that every decision you make impacts your chickens' quality of life. From predator-proofing to optimizing lighting and shade, each aspect contributes to creating a safe, comfortable home for your flock. With these foundational elements in place, you're ready to explore the next chapter: feeding and nutrition essentials. This new knowledge will further enhance your ability to care for and enjoy these delightful creatures.

CHAPTER 3
FEEDING AND NUTRITION ESSENTIALS

UNDERSTANDING CHICKEN DIETARY NEEDS AND OPTIONS

Witnessing the shimmering plumage and sprightly demeanor of my neighbor's chickens serves as an everyday testament to how profoundly a balanced diet influences poultry vitality. Just as diverse, nutrient-rich meals cater to human health, chickens thrive on diets loaded with essential nutrients that fuel their growth and overall well-being. A well-constructed diet can specifically increase their lifespan, improve their resilience against diseases, and enhance their productivity, reflecting how indispensable good nutrition is to their maintenance and management.

At the heart of a healthy chicken diet is protein, a critical building block facilitating robust muscle and feather development. Protein-rich feeds ensure not just physical growth but also energy for daily activities. Chickens engage in constant movement throughout the day, pecking at the ground and flapping their wings. This activity necessitates a steady supply of protein to replenish and repair muscle tissue, thereby acting as the bedrock of their energetic lifestyle. Meanwhile, calcium is indispensable for robust eggs, fortifying eggshells to prevent the disappointment of fragile, thin-shelled eggs that compromise egg quality and safety. Without adequate calcium, hens may suffer from weak bones, as the body will redirect calcium reserves to shell formation, jeopardizing their skeletal structure in the process and thereby compromising long-term health.

Furthermore, vitamins A, D, and E serve critical roles akin to guardians who bolster the immune system, ensuring that daily threats from pathogens remain at bay. Vitamin A is particularly crucial for maintaining healthy eyesight and skin, influencing the iridescent sheen of their feathers, which is often an indicator of a well-maintained chicken. Vitamin D, often synthesized through sunlight exposure, enhances the uptake of calcium, thereby serving a dual purpose in maintaining egg integrity and skeletal health. Vitamin E acts as an antioxidant, neutralizing free radicals, thus helping mitigate oxidative stress in chickens' bodies, which is vital to preserving their overall vitality and longevity.

As juvenile chicks journey from hatching to full-fledged hens or roosters, their nutritional needs undergo significant transformations. Chicks initially thrive on starter feed, which is abundantly rich in protein to accommodate their rapid initial growth and feather development. Starter feeds are often formulated to be easily consumable, with the grains and protein pellets finely ground to ensure chicks, whose pecking abilities are still developing, can consume meals efficiently. This dietary stage lays the groundwork for enduring health, imprinting well-being that echoes throughout their lives. As maturity beckons, their diet transitions towards layer feeds predominated by higher calcium levels. This adaptation coincides with their burgeoning roles as egg layers, heralding their contribution to egg production, which could potentially grace family breakfasts with their offerings.

Exploring feed options reveals a spectrum of possibilities, each catering to different farming philosophies and objectives. Organic feeds appeal to those championing pesticide-free, non-GMO nutrition. Such feeds are often perceived to enhance the overall quality of eggs and meat due to the absence of synthetic chemicals, potentially reducing the health risks associated with these additives. Organic feed options include ingredients like whole grains, fish meal, kelp, and unrefined barley malt, integrating seamlessly into an organic course of living. However, these luxurious options often come with a heftier price tag and may not be as accessible throughout all regions, thereby posing a dilemma for budget-conscious poultry keepers.

In contrast, commercial feeds offer a more approachable solution, usually fortified with a comprehensive array of essential minerals that promise balanced nutrition. Their affordability and broad availability make them particularly attractive, though some contain additives that certain keepers might prefer to eschew. It is imperative for poultry raisers to carefully scrutinize ingredient labels to avoid unwanted components, particularly for those with commitments to ethical farming or natural diets.

For enthusiasts eager to delve into the intricacies of advanced feeding methods, fermented feed arises as an intriguing alternative. Grains soaked until fermentation optimize digestibility and enhance nutrient bioavailability, paving the way for improved assimilation. Fermented feeds cultivate beneficial gut flora, leading to better digestion and improved egg production outcomes. Though this method demands patience and prior planning, involving placing grains in water and allowing them to sit for several days while stirring occasionally, it rewards caretakers with observable enhancements in the health and vitality of the flock. Fermentation not only enhances the nutritional profile but also reduces phytic acid levels present in grains, which otherwise may hinder mineral absorption.

NUTRITIONAL CHECKLIST

- **Protein Sources**: Fish meal, soybean meal, mealworms, peanuts
- **Calcium Supplements**: Crushed oyster shells or limestone
- **Vitamins**: A (carrots), D (sunlight exposure), E (spinach, nuts, and seeds)
- **Fiber Options**: Alfalfa hay, chia seeds, sunflower seeds

Navigating through the vast array of nutritional choices is both an exploration and a learning journey, considering the distinct nature and needs of every flock. Flock dynamics, including its size, age distribution, and breed, significantly influence the type of feed that best matches their needs. What ensures peak health for one coop may not necessarily suit another. Hence, maintaining sharp observational skills—attuning oneself to behavioral indicators like feather condition, activity levels, and egg quality—combined with making gradual adjustments based on chickens' responses to various feeds will guide you towards the most beneficial dietary balance, unlocking your flock's potential to sustain and thrive.

CHOOSING THE RIGHT FEED AND SUPPLEMENTS

Selecting the optimal feed is a nuanced process hinged significantly on the specific goals and compositions of your flock. For egg-laying hens, layer pellets emerge as the ideal nutritional blend. These pellets are crafted to combine ample protein with necessary calcium to support consistent egg production while maintaining overall health. The carefully calculated balance ensures hens remain in prime health, avoiding conditions such as osteoporosis. Conversely, broiler feed is apt for chickens raised for meat, possessing higher energy content to facilitate their rapid growth. This feed often includes a blend higher in corn and soybean meal—components renowned for their energy-yielding properties.

Understanding the nuances of a nutritional pyramid is crucial for choosing feed that aligns with your chicken-keeping goals. This pyramid, similar to those used in human nutrition, emphasizes the balanced integration of proteins, grains, minerals, and vitamins, each layer contributing to the overall health and productivity of your chickens. At its base, grains provide essential carbohydrates for energy, while the next layer up, proteins, support muscle and feather development. Above proteins, vitamins and minerals form a critical layer, ensuring immune health, strong eggshells, and robust skeletal structures. The apex of the pyramid focuses on supplements such as calcium and grit, enhancing diet specifics where standard feed may fall short. By familiarizing yourself with this structured approach, you can more effectively assess whether a specific feed formulation meets the comprehensive dietary requirements of your flock, ensuring they receive the balanced nutrition necessary for optimal health and egg production.

Location-specific availability of these feeds necessitates interaction with local farm stores or co-ops, where one can not only discover varied feeding options but also benefit from the guidance of fellow chicken enthusiasts. These resources offer invaluable insights and advice tailored to your locality's specific farming ecosystem. Engaging with the local farming community cultivates a network of shared knowledge, experiences, and collective purchasing power.

Supplements are essential dietary aids, acting as crucial bolsters where primary feed falls short of comprehensive nutrition. For instance, oyster shells serve as a vital calcium boost to bolster strong eggshell formation and combat issues like egg binding. Chickens are susceptible to a range of conditions if their diet is not adjusted for their specific physiological demands, making careful observation and strategic supplement incorporation essential. Likewise, grit is another necessary supplement, particularly for chickens lacking natural access to gritty soil, such as non-free-range birds. This gritty material aids in the grinding of food inside the gizzard, an essential process for digestion due to chickens' lack of teeth.

Seasonal changes and climate variables demand considerable dietary adjustments for chickens. When winter's icy grasp encroaches, their energy needs increase to sustain body warmth, prompting shifts in feeding strategy. Corn, with its high calorific content, becomes akin to a warming coat against winter's chill, stabilizing their metabolism amid frigid temperatures. Conversely, during the peak summer when the sun blazes with ferocity, maintaining hydration is paramount. Offering juicy fruits or hydration-rich vegetables like watermelon furnishes a welcome reprieve from the heat, helping chickens regulate body temperature while guaranteeing essential hydration levels. Such considerations ensure the flock's comfort and productivity are maintained regardless of climatic upheavals.

Understanding the cyclical nature of the chickens' breeding and laying patterns can further enhance the adjustment of their feed. During peak laying seasons, a higher concentration of calcium will be necessary, while during molting phases, increased protein and supplementary vitamins assist in feather regrowth and maintaining overall health.

Ensuring quality in chosen feed brands can significantly uplift poultry health and productivity. Brands such as Purina Layena and Nutrena NatureWise have gained recognition for their specialized formulations supporting poultry vitality. Purina Layena is precisely engineered for optimal egg production, while Nutrena NatureWise offers a range accommodating various poultry needs with its balanced formulations. These branded feeds are regularly tested to uphold high-quality standards, ensuring chickens receive adequate nutrition for maximal health benefits. Prioritizing these brands and maintaining consistency in feed quality ensures sustained health and productivity in your flock.

Reducing feeding costs without diminishing quality demands strategic planning and savviness. Purchasing in bulk offers discounts that accumulate into significant savings over time. Many local feed stores promote bulk purchases, occasionally accompanied by loyalty programs for additional cost benefits. Moreover, crafting homemade feed mixes tailored to your flock's requirements offers further opportunities for cost-saving, enabling you to purchase individual feed components economically and fashion a custom blend that meets nutritional requisites without relying heavily on commercial feed products.

Mitigating waste stands crucial for managing feeding costs effectively. Adequate storage solutions, such as airtight containers, ward off spoilage and pest infestation—a vital countermeasure considering most grains' susceptibility to these risks. Feed troughs designed to minimize spillage help conserve food, ensuring maximum nutrition reaches the chickens rather than being scattered and wasted. Implementing these storage practices prevents rodents and other pests, helping maintain nutritional integrity and lowering potential health risks.

Incorporating kitchen scraps provides an enriching nutritional variation while decreasing food waste. Foods like vegetable peelings, fruit scraps, and leftover grains add dietary diversity. Nevertheless, it is essential to avoid harmful foods such as avocados or onions. Removing uneaten leftovers promptly and chopping larger scraps prevent spoilage and deters pests, maintaining a clean, safe feeding environment.

Adjusting dietary plans for seasonal changes is imperative. Ensuring proteins are available for feather insulation during the colder months, while hydration-rich foods take precedence during summer's heatwaves, aligns diet with environmental demands. Reactive adjustments based on sudden climatic changes could involve increasing energy-rich foods during cold spells or offering cooling treats during heatwaves.

Ensuring clean and uninterrupted access to water is a non-negotiable requirement for maintaining flock health. Hydration is critical for digestion, egg production, and overall vitality. Automatic waterers and heated water bowls offer practical solutions to provide constant water supply even during freezing conditions, thus eliminating the need for frequent manual intervention during harsh weather conditions.

These strategies form the foundation of a year-round nutrition-focused approach to poultry care, establishing a sustainable path towards nurturing a thriving, robust flock. Consistency in application ensures the flock remains resilient and productive, aligned with scheduled husbandry cycles that mirror the natural environment chickens thrive in.

MANAGING FEEDING COSTS WITHOUT COMPROMISING QUALITY

Upon commencing their journey into chicken raising, many have been taken aback by how quickly feed expenses could accumulate. Establishing cost-effective practices should become a cornerstone of an efficient management strategy, ensuring the flock's health without exerting excessive financial pressure. The realization that small decisions accumulate to bear significant financial implications will inform one's drive towards cost-conscious feeding practices. For many, embracing bulk purchasing discounts have emerged as a primary strategy for substantial savings. Collaborating with local cooperatives or partnering with suppliers often yields opportunities for alluring bulk discounts. This strategy can be more effective when organizing with other local chicken keepers, allowing small-scale buyers to access bulk rates traditionally reserved for larger operations. However, these purchases necessitate verifying the adequacy of storage facilities to manage larger quantities effectively without encountering spoilage or infestation issues.

Enhancing dietary strategies through homemade feed blends is a beneficial practice embraced by experienced chicken keepers. This avenue permits tailoring feed to align precisely with nutritional needs, offering greater control and potentially boosting cost efficiency. Exploring diverse recipes that dovetail with chickens' nutritional mandates is advisable, offering flexibility in adapting feed compositions to suit seasonal or breed-specific needs. Opt for grains such as oats or barley as the foundational feed base, enriching it with protein fortification derived from fish or soybean meals. Such hands-on approach enables control over ingredient quality and makes nutritional customization accessible, reducing dependency on generic commercial products. This do-it-yourself approach provides transparency in ingredient selection and can economically align with nutritional aspirations comparably better than commercial feed reliance.

Regularly tracking your expenses for feed and supplements is key to managing your budget effectively. By using budgeting tools or templates, you can keep a close eye on your spending, offering a comprehensive overview of your costs. Recording your expenses for storage and maintenance supplies also helps reveal how resource use varies with the seasons, pinpointing opportunities for budget refinement or essential investments. This diligent financial tracking is crucial for optimizing your financial management, securing the long-term sustainability of your chicken-raising venture.

BUDGETING TEMPLATE

- **Monthly Feed Expenses**: Record bulk purchases and individual bags.
- **Supplement Costs**: Catalog additional items such as grit or oyster shells.
- **Storage Expenses**: Factor in costs for airtight storage containers and pest control measures.

- **Unexpected Costs**: Keep track of any additional unexpected expenses such as veterinary care or emergency feed solutions.

Utilizing these financial management templates is crucial for making well-informed decisions that prioritize cost-efficiency without skimping on your flock's nutrition. The core of effective budgeting is to sustain practices that ensure quality care without overspending, thereby promoting the well-being of your flock within a manageable budget. This strategy safeguards the health of your chickens while also streamlining expenses to reinforce the economic foundation of your chicken-keeping venture.

Building networks within local communities brings you closer to fellow chicken enthusiasts, fostering collaboration in bulk purchases or shared experiences. Such camaraderie nurtures a vibrant, innovative atmosphere that enriches your poultry-keeping journey, potentially leading to joint endeavors in feed production or shared success stories in devising cost-saving strategies. The knowledge exchange in local farming groups often highlights inventive solutions one might not have independently considered, offering sustainable ways to reduce costs without undermining the nutritional quality delivered to the flock.

Embracing these strategies not only conserves resources and enriches the chicken-raising experience but also bolsters flock care quality, devoid of unnecessary financial burdens. By leveraging collaborative knowledge, embracing strategic planning, and refining feed selection, nutritional needs can be met meticulously while respecting budgetary constraints, ensuring the flock—and its caretakers—prosper sustainably.

CHAPTER 4
HEALTH AND WELLNESS OF YOUR FLOCK

RECOGNIZING AND TREATING COMMON CHICKEN DISEASES

T he first time I observed one of my neighbor's cherished hens appearing unwell, it made a lasting impression on me, highlighting the deep care and responsibility involved in raising chickens. The signs of her ailment were unmistakable: her previously lively comb had become droopy, resembling a wilting flower, a clear sign she was not her usual spirited self. This incident prompted me to delve into the various diseases that can affect poultry, each presenting a unique threat to their health. Among these concerns, respiratory infections such as Infectious Bronchitis beckon immediate attention. This particular illness weaves through flocks with unnerving speed, characterized by symptoms such as persistent sneezing and a decline in egg production, which could spell economic loss for a farmer relying on egg sales. Its rapid spread underscores the necessity for early intervention.

Equally distressing are parasitic threats like coccidiosis, especially notorious among the avian community. Affecting primarily young chicks, coccidiosis manifests through diarrhea and stunted growth, emphasizing the criticality of vigilance from the onset of life. A proactive approach in detecting illness involves keen observation, honing skills to detect the slightest shifts—a despondent demeanor, avoidance of the flock, or a lack of interest in regular feed. These signs could be precursor signals of underlying health issues that demand prompt response.

Effective treatment is the backbone of flock health, providing relief and recovery when faced with adversities. Antibiotics, judiciously administered under strict veterinary supervision, can be pivotal against bacterial ailments. However, their careful and controlled use is critical to prevent undue resistance. Exploring complementary treatments, such as natural remedies, can serve as an adjunct to traditional medical care. Garlic, known for its mild antibacterial properties, can be introduced by crushing it into the water supply, potentially forestalling minor infections. Nonetheless, it's imperative to recognize when these home remedies fall short and professional medical intervention must take precedence, ensuring the provision of the best possible care to your flock.

Beyond treatment, veterinarians play a vital role in preventive health measures, guiding you to preserve the general well-being of your flock. Building rapport with a vet seasoned in poultry care could hold immense value, ensuring the availability of advice for varied situations. Investing time in crafting a health protocol tailored to your flock's unique needs, along with conducting regular veterinary screenings, can preempt many potential health issues.

ADDITIONAL SYMPTOM INDICATORS

These newly introduced symptoms could also offer a broader view:

Symptom	Possible Cause	Action
Lethargy	Viral Infection or Nutritional Deficit	Conduct blood tests, reassess diet, consult a vet
Sudden Weight Loss	Worm Infestation or Lowered Immunity	Initiate deworming, supplement vitamins, seek vet
Limping	Injury or Infectious Arthritis	Isolate and observe, provide anti-inflammatory care

Consistent monitoring reinforces your ability to swiftly identify disorders, forming the core of an effective response strategy. This knowledge empowers reaction, transforming potential health challenges into educational journeys, fostering a nimble, prepared approach toward chicken keeping.

EFFECTIVE PARASITE CONTROL AND PREVENTION

When first venturing into the realm of parasites, it can seem daunting; however, understanding these stealthy invaders is crucial to safeguarding your flock. Both internal and external parasites threaten chicken health, with lice being one of the smallest yet most troublesome. These minuscule insects may inhabit the feathers, causing discomfort and potential bald patches due to persistent scratching. Mites, elusive by nature, make their moves in the stillness of the night, eschewing daylight and latching onto their hosts only under the guise of darkness. Among internal parasites, notorious worms like roundworms and tapeworms wreak havoc on a chicken's internal system, siphoning essential nutrients and causing weight loss if untreated.

A strategic prevention plan is crucial to avoiding infestations altogether. The foundation begins with cleanliness; maintaining a tidy coop environment becomes your first line of defense. Employing diatomaceous earth—a powder derived from ancient deposits of fossilized algae—acts as a natural deterrent and disrupts parasite life cycles. Consider it akin to constructing a stronghold against invaders. Moreover, regular changes of bedding, as well as ensuring proper ventilation, create an inhospitable environment for these pests, keeping them at bay.

Addressing active parasite infestations necessitates a multifaceted approach to treatment. Utilizing topical solutions such as dusting powders and sprays can effectively target lice and mites, while a deworming schedule, as recommended by a veterinarian, ensures internal parasites are also managed. Consulting with poultry health experts can enhance your approach, offering proven strategies to safeguard your flock's health. Additionally, incorporating natural deterrents, such as oregano oil, offers a chemical-free method to repel pests. A few drops in their water can serve as an effective, nature-inspired preventative measure, blending the best of traditional and holistic parasite control techniques.

A finely honed checklist aids in the ongoing battle against these intruders:

ADVANCED PARASITE CONTROL CHECKLIST

- Implement rotational grazing to break parasite life cycles.
- Incorporate antiparasitic plants such as wormwood within run space.
- Designate ample dust bathing areas to promote natural cleaning.

Step	Action
Coop Maintenance	Weekly deep clean, replace or sanitize bedding
Birds Inspection	Inspect for signs of parasites bi-weekly, focusing on vents and feathers
Herbal Integrations	Plant herbs like thyme and lemon balm near the coop
Natural Treatments	Integrate feeds with natural antiparasitic properties

With vigilance and preparedness, even the tiniest of threats can be neutralized, maintaining your chickens' thriving condition. By combining thorough maintenance, proactive treatment, and preventive measures, you empower your flock with the tools needed to live vibrantly. A pest-free coop is a testament to diligence, enabling you to reclaim serenity knowing your efforts yield a robust and lively flock.

VACCINATION BASICS FOR A HEALTHY FLOCK

Initially, the concept of vaccinations was overwhelming and somewhat unclear. However, it quickly became evident that vaccinations are essential for safeguarding the flock's health. Vaccinations erect an essential barrier against formidable diseases. For instance, outbreaks of Newcastle disease, a highly contagious respiratory virus, can wreak havoc throughout a flock. Vaccinating chickens greatly diminishes its spread, so their overall health is sustained, which dovetails with the owner's peace of mind. Marek's disease is another complex challenge—known for causing tumors and paralysis without warning. Vaccinations aimed at this disease significantly curtail its reach, enhancing the flock's collective resilience.

Certain vaccines hold prominence in the backyard chicken realm. The Fowlpox vaccine stands as a bulwark against a virus notorious for grotesque lesions and formidable health detriments. Similarly, vaccinating against Avian Encephalomyelitis shelters young chicks from a viral disease leading to tremors and motor dysfunction. Incorporating these vaccinations into your flock's health plan efficiently shields against potential catastrophe, nurturing a healthy, thriving group of chickens.

Timing is a critical element in the vaccine regimen. Initial vaccination doses are typically administered to chicks when they're only a few days old to establish foundational immunity. Booster schedules are essential to maintaining continuous protection against evolving pathogenic threats. Regularly monitoring these schedules and adhering to follow-up doses helps achieve optimal long-term efficacy.

Multiple vaccination methods afford flexibility based on specific requirements. Injectable vaccines deliver precise dosages, albeit necessitating the handling of each individual bird. Alternatively, water-based vaccines offer a less invasive option, ideal for larger flocks, although they require exact mixing and meticulous application to ensure thorough dissemination.

Compliance with vaccine packaging instructions is essential for effectiveness and safety, necessitating proper storage and handling. These guidelines maximize vaccine potential, ensuring it remains a robust tool in maintaining flock health.

Vaccinations embody a strategic investment to bolster productivity and diminish worries over poultry health, preserving peace of mind and yielding prosperous chicken endeavors.

Sourcing vaccines might often be challenging for smaller operations as commercial solutions cater predominantly to large-scale enterprises. Acquiring pre-vaccinated birds through hatcheries circumvents logistical concerns, providing a seamless and effective alternative.

While integrating vaccinations within your management plan is crucial, they are just one tier of comprehensive flock care. Paired with stringent hygiene practices and routine health audits, vaccinations become part of a holistic strategy leading to optimal poultry well-being.

Each proactive step warrants an environment conducive to thriving chickens. Armed with a vaccine regimen, you will be equipped to robustly defend against potential challenges, cultivating a vibrant and dynamic flock.

UNDERSTANDING CHICKEN BEHAVIOR AND SOCIAL DYNAMICS

Chickens inhabit a world characterized by richness in social dynamics. Chickens establish a social hierarchy, commonly known as the "pecking order," that is crucial for flock harmony. This system determines access to vital resources such as food and preferred roosting spots. Seeming from the outside like a cacophony, this precise order is fundamental in maintaining intra-flock harmony. The introduction of new hens disrupts this balance, necessitating rank establishments and triggering necessary scuffles as each bird finds its footing. Although this may seem aggressive, it's a natural integration process that stabilizes once order restores within the hierarchy.

Foraging and dust bathing, more than mere routine activities, are essential for chicken satisfaction. Foraging encompasses more than aimless pecking; it's an exploratory activity tapping into their instincts while offering mental and physical stimulation. Dust bathing operates as a form of self-cleaning, an effective means to combat parasites, and a source of evident joy. Watching them whirl in delight within the dust privileges the onlooker with a glimpse of their innate happiness and contributes to health preservation.

The presence of a dominant hen or rooster sheds light on the intricacies governing social dynamics. Often more assertive rather than the largest, this bird naturally assumes leadership, guiding the flock steadily between resources and shelter, fostering a sense of tranquility. Nonetheless, vigilant monitoring is crucial to intercept aggression that sometimes ensues from excessive authority. Instances of bullying must be addressed promptly, ensuring harmony and welfare across the flock spectrum.

Sensitivity to behavioral fluctuations, such as sudden withdrawal, often signals underlying discomfort. Environmental precursors, such as changes in weather, often dictate shifts in behavior patterns, serving as vital predictive cues. Strategic redirection via new perches or introducing hanging vegetables as interactive distractions assists in curbing undesired conduct. Creating an environment filled with engaging activities reduces behaviors caused by boredom, leading to healthier interactions among the flock.

Keeping a detailed behavior journal aids in elevating understanding and documentation of flock interactions:

BEHAVIOR JOURNAL PROMPT

- Daily interactions and observations.
- Noting any behavioral shifts or changes in dynamics.
- Recording intervention attempts and corresponding results.

Maintaining a detailed record offers valuable insights and serves as a resourceful guide for fine-tuning social dynamics approaches. Enhanced comprehension of chicken behavior solidifies your management repertoire while intensifying your intrinsic connection with the flock.

MANAGING STRESS AND ENSURING FLOCK HARMONY

Chickens, like all beings, endure stress, impacting both health and productivity. Spotting potential stress inducers is integral to promoting a serene living environment. Overcrowding, a prominent catalyst, incites competition for essentials such as food and water, drawing parallels to our own discomforts amidst restricted resources. Ensuring ample space both within the coop and the run alleviates such pressures considerably. Additionally, abrupt environmental shifts, such as relocating the coop or altering feeding schedules, can disturb chickens, disrupting the tranquility of their environment.

Mitigating stress requires thoughtful strategies, with gradual introductions of new members easing potential tension. Allowing visual contact prior to cohabitation smooths subsequent transitions. Reliable care routines contribute to flock security by providing predictability and trust, dissolving uncertainties tied with unpredictability.

Fostering harmony necessitates methodology. Installing

multiple feeders reduces resource competition akin to distributing buffet stations, enhancing accessibility during crowded gatherings. Providing plentiful roosting space ensures peaceful rest, mitigating conflicts over prime spots. Proper arrangements promise comfort for all birds, reducing disputes substantially.

Implementing minor adjustments translates into significant strides towards flock peace. During the integration of newcomers, initial chaos might appear formidable, yet strategic interventions like temporary barriers or adjustments to yard access encourage gradual reconciliation, attaining harmony over time. Expanding physical space with portable fencing reframes dynamics, alleviating pecking disputes resulting from spatial constraints.

In addition to physical adjustments, mental stimulation plays a crucial role in maintaining flock harmony. Introducing activities that engage their minds and bodies significantly reduces aggressive behavior, leading to a more peaceful group. Providing hanging vegetables or creating designated dust bathing areas adds enrichment to their daily routine, akin to the comfort toys offer during unfavorable weather conditions. A mentally engaged and happy flock is often more productive, enhancing the overall health and harmony within the coop.

Managing stress and maintaining harmony within your flock requires a delicate equilibrium. Yet, each improvement you make significantly contributes to the tranquility of your chickens. By observing their interactions closely, you can tailor strategies that specifically cater to their needs, fostering a peaceful environment. Embracing flexible approaches and being willing to adjust your methods encourages a harmonious balance, ensuring your flock thrives in a stress-free setting.

To create a tranquil haven for your chickens, it's essential to grasp and cater to their needs thoughtfully and deliberately. Recognizing and reducing stress factors, coupled with nurturing a harmonious environment, builds a strong foundation for a thriving chicken community.

HOLISTIC APPROACHES TO CHICKEN HEALTH

Holistic health adds a layer of vitality to chicken care, prioritizing their overall well-being to enhance resilience and energy. Incorporating herbal supplements like echinacea and elderberry into their feed or water leverages their immune-boosting benefits.

Similarly, incorporating probiotics plays a crucial role in fortifying gastrointestinal health by introducing a diverse community of beneficial bacteria. This introduction supports a more balanced digestive ecosystem, facilitating enhanced nutrient absorption and bolstering the immune system. Available in forms such as powders, liquids, or directly infused into feed, these supplements can be easily integrated into your chickens' daily diet. The consistent use of probiotics not only contributes to improved digestive health but also promotes overall well-being, leading to happier, more content chickens. By prioritizing gut health through the strategic use of probiotics, you're investing in a foundational aspect of your flock's holistic care, ensuring they maintain optimal health and contentment.

Cultivating a healthy environment remains essential. Sunlight exposure facilitates vitamin D synthesis, invaluable for bone strength and robust egg production. Strategically positioning coops or incorporating windows maximizes sunlight access. Alongside light, herbs like mint and lavender exude aromatic charm while concurrently repelling insects, crafting an inviting habitat.

Promoting natural foraging behaviors allows chickens to enhance their diet beyond what standard feed offers. Their inherent curiosity drives them to peck, scratch, and graze, uncovering a variety of nutrients not found in their usual meals. This engagement with their environment not only fulfills their nutritional needs through a broader spectrum of resources but also activates their natural instincts, adding richness to their daily routines.

Balancing holistic approaches with traditional care methods integrates a comprehensive health strategy. Genuine symbiosis between vaccinations and natural remedies offers a reliable defense, fused with enriched resilience.

Embracing holistic care encompasses a blend of preventative and curative approaches, setting the foundation for a thriving flock. Picture your chickens basking in the warmth of the sun, their feathers shimmering with health, embodying the essence of vitality. This approach nurtures an environment that allows your chickens to thrive, promoting their well-being and ensuring a life of vigor and longevity.

In conclusion, holistic health transcends addressing surface challenges, instilling environmental support fostering physical and mental wellness. The next chapter will delve deeper into the egg production process, journeying from coop life to kitchen transformations.

MAKE A DIFFERENCE WITH YOUR REVIEW

UNLOCK THE POWER OF GENEROSITY

"The best way to find yourself is to lose yourself in the service of others." – Mahatma Gandhi

People who give without expecting anything back often feel the happiest. So let's do something kind together!

Would you help someone just like you—curious about raising chickens but unsure where to start?

My goal is to make raising chickens simple, fun, and rewarding for everyone.

But to reach more people, I need a little help.

Most folks choose books based on reviews. So I'm asking you to help a fellow chicken-keeper by leaving a quick review.

It's totally free and takes just a minute—but it could make a big difference in someone's chicken journey. Your review might help...

...one more family gather fresh eggs from their backyard.

...one more kid learn where food comes from.

...one more person enjoy the peace that comes from caring for animals.

...one more beginner feel confident to start.

...one more dream of a simple, happy life come true.

To leave a review, just scan the QR code or visit this link:
https://www.amazon.com/review/review-your-purchases/?asin=B0F34TZ1RK

If you enjoy helping others, you're my kind of person.

Thank you so much for your support—it truly means the world.

— Avery Sage

CHAPTER 5
EGG PRODUCTION AND QUALITY

FACTORS INFLUENCING EGG PRODUCTION IN CHICKENS

Envision the comfort of your morning routine, cradling a warm cup of coffee as you admire a basket filled with fresh eggs gathered from your own backyard hens. This moment of tranquility highlights the deep bond shared between you and your flock. To sustain these quiet moments, understanding the myriad factors influencing egg production is crucial. For example, environmental conditions play a pivotal role. As daylight wanes with the onset of winter, egg production tends to decline. Chickens thrive on 14 to 16 hours of light a day to sustain peak laying performance. To counteract nature's tempered light during darker months, think about integrating supplemental lighting. Employing artificial lighting wisely can recreate the effect of longer daylight hours, encouraging your flock to sustain their egg-laying rates even through the shorter days of winter.

However, not just artificial lighting supports productivity; the strategic placement of windows in the coop can harness natural light, creating a more environmentally conscious approach to augmenting light exposure. This dual approach blends the benefits of technology with nature's rhythms, fostering an environment where hens feel aligned with natural cycles despite seasonal changes.

Temperature fluctuations rank high among variables affecting chicken productivity. In sweltering heat, hens expend vital energy regulating body temperature, while in biting cold, their efforts shift toward generating warmth over egg production. It's essential to create a controlled environment that minimizes these extremes. Implementing insulated coops, installing fans for ventilation, and erecting shade structures can regulate temperatures favorably, ensuring your hens remain comfortable, thus optimizing their laying cycles.

Beyond these solutions, the importance of microclimates within a coop could be explored further. By positioning water sources and ventilation openings strategically, multiple temperature zones can be established within a single coop. This variety offers hens choices, enabling them to seek out the most comfortable conditions to maximize their productivity.

Genetics is another cornerstone influencing output. Egg-laying breeds, like the highly efficient Leghorns, showcase remarkable genetic capabilities, yielding upwards of 300 eggs each year. Your selection of breeds inevitably decides the yield—heritage breeds like Orpingtons or Sussex may produce fewer eggs annually but offer exceptional qualities like sturdiness and amiable temperaments. A comprehensive understanding of your breed's genetic makeup guides realistic expectations, ensuring that goals align with biological possibilities.

Expanding upon breed considerations, the genetic inclination towards winter laying is an additional aspect worth examining. Certain breeds inherently manage colder conditions better, continuing to lay throughout the winter with minimal encouragement. This genetic insight can help orient new flock introductions based on regional climatic patterns—choosing a breed that inherently fits your environment maximizes success.

Age is a significant determinant of egg production. The prime years, typically between one and two years old, showcase a hen's peak laying potential, often leading to abundant egg collection. Contrarily, as hens age beyond three years, egg production diminishes progressively, a shift reflecting their physiological tilt towards sustenance and corporeal maintenance over reproduction. Strategically planning the timeline for introducing new birds into your flock, therefore, plays a key role in ensuring a continuous and balanced supply of eggs.

Even more, incorporating temporary substitutions into the flock, such as younger hens or pullets, can mitigate the natural production dip as older hens taper off. This consideration ensures that production levels remain steady, as the younger birds begin to compensate for the reduced output from their elder counterparts.

Stress and health are substantial influencers on egg yield. Frequent predator sightings often disrupt your flock's routine, potentially halting their laying. Guaranteeing a secure coop and run is vital for shielding chickens, reducing fear-induced production disruptions.

Health complications, such as egg binding—a serious condition where an egg becomes stuck within the hen's reproductive tract—demand prompt and careful attention to avert critical health issues. Egg binding can be caused by a variety of factors, including nutritional deficiencies, particularly a lack of calcium, obesity, or genetic predispositions. Recognizing the signs of this condition early, such as a hen straining without producing an egg, lethargy, or a swollen abdomen, is crucial for timely intervention. Immediate measures might include gently warming the hen to relax her muscles or consulting a veterinarian for professional advice. Preventative strategies, focusing on proper nutrition, regular exercise, and stress reduction, play a key role in minimizing the risk of egg binding and ensuring the health and productivity of your flock.

INTERACTIVE ELEMENT: EGG PRODUCTION JOURNAL

Consider launching an egg production journal. Document environmental factors, daylight variations, and stressors impacting your flock. Regular entries serve as diagnostic tools, unveiling patterns and offering insights into laying trends. Proactive management through journaling enables you to make informed decisions, enhancing egg production. Recognizing alterations in laying patterns early can illuminate the need for interventions, fostering adjustments that optimize productivity. This simple yet powerful practice acts as a guide for sustaining balance within the flock, ensuring that each egg collected mirrors the efforts invested in creating an ideal living environment for your hens.

Incorporating various tracks, such as egg size and weight over time, can further enrich this journal, offering comprehensive insight into both immediate conditions and longer-term trends. The journal not only acts as a day-to-day log but evolves into a valuable resource for fine-tuning hen management strategies.

ENHANCING EGG QUALITY WITH PROPER NUTRITION

Cracking open an egg to reveal a vibrant yolk housed in a shell of pristine firmness speaks volumes of the nutritional care provided to your chickens. Essential nutrients play pivotal roles in achieving such quality, beginning with calcium—indispensable for forming strong shells. Insufficient calcium intake often results in fragile, easily cracked shells. Offering calcium supplements like crushed oyster shells or limestone fortifies shell structure, creating a preventive barrier against breakage. Omega-3 fatty acids are equally vital for augmenting yolk quality. Common in ingredients like flaxseed or fish meal, these beneficial fats enrich the nutritional value of your eggs, enhancing both their visual appeal and health benefits.

Moreover, vitamin D is crucial in calcium absorption, and ensuring your flock has access to natural sunlight can exponentially increase egg quality. For coops situated in areas with limited sunlight exposure, vitamin D supplements can fill the gap, sustaining the interdependent relationship between calcium structures and vitamin effectiveness.

A balanced diet is crucial for maintaining exceptional egg quality. Layer pellets, rich in proteins, vitamins, and essential minerals, should form the foundation of this diet, tailored specifically for laying hens. The monotony of pellets can be expanded upon by integrating fresh greens like kale or spinach. These additions not only diversify dietary intake but also infuse eggs with enhanced color and flavor, ensuring a broad spectrum of nutrients reaches your flock. Such variety promotes robust health, enhancing productivity.

Natural additives present another layer of nutritional fortification, readily elevating egg quality. Flaxseed, abundant in omega-3s, boosts both yolk color and content. Incorporating flaxseed into the diet manifests visually and health-wise, producing golden yolks that stand as testaments to nutritional diligence. Apple cider vinegar, an unsung hero in poultry nutrition, aids digestion and fortifies gut health. Adding a splash to your hens' water improves overall nutrient absorption efficiency, translating into tangible benefits reflected in egg quality.

Dive deeper into the idea of homemade feed options—an innovative approach that satisfies specific dietary needs while minimizing food waste. Scraps such as citrus peels and small amounts of garlic can offer antioxidative benefits, subtly complementing standard poultry feeds. This judicious practice reduces waste and contributes to a closed-loop system in backyard husbandry.

Crafting homemade feed is an alternative that meets specific dietary needs while reducing waste. Kitchen scraps find renewed purpose in this approach—remnants like carrot tops and lettuce leaves become nutritious supplements when chopped and mixed with grains such as corn or oats. This practice transforms potential waste into fresh dietary additions, supporting your flock's health.

Custom feed mixes offer further tailoring opportunities. Begin with base grains—corn or wheat blends well—supplementing with protein-rich ingredients like soybean or fish meal. Seeds add nutritional and entertainment value, enriching your flock's diet. This approach empowers you to dictate precise nutritional inputs, maximizing health and egg quality potential.

HOMEMADE FEED RECIPE

A straightforward recipe might involve:

1. **Base Grains:** 50% corn, 20% wheat
2. **Proteins:** 10% soybean meal, 10% fish meal
3. **Additives:** 5% flaxseed for omega-3 enrichment, 5% sunflower seeds for taste and texture

Thoroughly mix these elements and store in airtight containers to preserve freshness. Experiment with portion sizes to accommodate fluctuating seasonal needs, understanding that dietary tweaks might be necessary to align with changing production phases.

Balancing dietary needs while enhancing egg quality invites creativity into chicken husbandry. A diet rich in calcium and omega-3s fortifies eggshells and imbues yolks with vibrance. Supplementing with fresh greens and natural additives nurtures overall health and laying proficiency. Homemade feeds utilizing kitchen scraps not only minimize waste but also grant control over nutritional dynamics. Adjusting these diets seasonally or based on specific needs ensures chickens remain healthy and productive year-round. This careful curation transforms basic ingredients into enriching meals, culminating in eggs of superior quality—a nourishing delight for both chickens and their keepers.

TROUBLESHOOTING COMMON EGG PRODUCTION ISSUES

The disappointment of fewer-than-usual eggs or eggs with unusual shapes and textures is a shared frustration among chicken keepers. Diagnosing the core issues requires a blend of observation and informed detective work. Nutritional deficits often lie at the heart of egg production challenges. Inadequate calcium intake, for example, manifests in thin or soft eggshells. Similarly, a deficiency in essential nutrients can lead to a downturn in egg count.

Delving deeper into the aspect of feed palatability, we uncover additional complexities. Changes in the flavor profile of feed, whether from ingredient degradation or shifts in the quality of feed components, can impact hens' willingness to eat adequately. This reluctance can negatively influence egg production. By conducting regular taste tests of feed batches, any issues with feed attractiveness can be identified early, allowing for swift corrective measures.

Mitigating stress factors is essential for maintaining egg production. The presence of predators or sudden changes in their environment can cause hens to stop laying as part of a natural survival response. To counteract this, reinforcing coop security, providing a diet rich in essential nutrients, and effectively managing any stressors are key measures.

Real-life experiences provide invaluable insights. Consider a seasoned chicken keeper who faced a calcium deficiency in her birds, resulting in fragile shells. By augmenting their diet with oyster shell supplements and ensuring a nutritionally balanced feed, production levels improved remarkably within weeks. Another scenario involved a drop in egg numbers following a hawk sighting. Installing reflective deterrents and providing enclosed shelters restored the flock's comfort and laying patterns, illustrating how addressing stressors impacts productivity.

Understanding your flock's needs, observing behavioral shifts, and implementing targeted solutions form the core of effective egg production troubleshooting. Each issue resolved adds to your skill set, enhancing your flock's well-being. Although challenges exist, the satisfaction gleaned from overcoming them is immense, reinforcing the value of care invested in chicken keeping.

EFFICIENT EGG COLLECTION AND STORAGE TECHNIQUES

Walking into your coop to find a nest of pristine eggs is a delightful testament to your hens' labor. Effective egg collection and storage not only maximizes freshness but enhances enjoyment. Aim to collect eggs daily, preferably in the morning, when hens complete their laying. Regular collection mitigates spoilage risks and keeps eggs clean, deterring pests and ensuring quality.

Handling eggs with care is essential to avoid cracks and extend their shelf life. After collection, the focus shifts to proper storage to ensure they remain fresh. Storing eggs at approximately 45°F is ideal for maintaining their quality. While refrigeration is the most effective method, storing eggs at a cool room temperature, away from direct sunlight and heat sources, can also preserve their freshness. Placing eggs in cartons with the pointed end down helps maintain the yolk's central position, keeping the egg's structure and flavor intact.

Exploring methods to verify egg freshness is equally important for extending their utility. The float test is a simple yet effective technique that identifies the accumulation of gases in eggs, indicative of aging. By placing an egg in water, its buoyancy reveals its freshness—fresh eggs sink, while older ones may float. This rapid assessment tool helps in reducing waste, allowing for the early identification of eggs that are past their prime.

Traditional preservation techniques, such as water glassing, offer a nod to the age-old methods of extending egg freshness without the need for refrigeration. By submerging eggs in a solution of lime water, a protective barrier against bacteria is formed, significantly prolonging the eggs' shelf life. When stored in this manner, eggs can remain fresh for up to three weeks, even when kept outside the fridge. Delving into these historical preservation methods connects us to the innovative approaches of past poultry keepers, who skillfully navigated the challenges of egg storage long before the advent of modern refrigeration.

Exploring modern preservation techniques, like freezing eggs after they are cracked and stored properly, extends their lifespan significantly. This approach enables efficient inventory control, guaranteeing that no egg is wasted during times of high production when the rate of laying surpasses immediate consumption demands.

Whether rooted in ancestral lime water baths or facilitated by contemporary refrigerators, ensuring egg freshness reflects a commitment to quality. Appreciating every egg laid, knowing each was collected and stored with intention, uplifts the broader experience of poultry husbandry. Effective handling cultivates a sense of connection to the food cycle, affirming the bountiful returns derived from disciplined practices.

UNDERSTANDING THE MOLTING PROCESS AND ITS IMPACT

The sight of feathers strewn across your backyard may alarm initially, but it signifies the natural molting process. Molting is a regenerative cycle where chickens shed older feathers for new ones, impacting egg production. During molting, hens redirect energy towards feather growth, temporarily pausing egg-laying. Their bedraggled appearance belies the dynamic changes occurring beneath, as they prepare for renewal.

For some enthusiastic poultry keepers, understanding the biological nuances behind molting adds depth to compassionate care. Distinguishing between partial and full molts can alter your flock's management strategy. Partial molts, often minor and less demanding, contrast substantially with the extensive demands of a full molt, necessitating variable dietary adjustments.

A nutritionally demanding phase, molting calls for increased protein intake, akin to athletes in training. High-protein feeds, accented by nutritious snacks such as mealworms, support regrowth. Vitamins like A and E also contribute to recovery, promoting tissue health.

Behavioral changes accompany molting, too. Reduced activity levels allow hens to conserve energy. Temporary shifts in flock hierarchy are common as some birds, vulnerable due to feather loss, maneuver social structures. A caring approach involves adjusting the environment to meet nutritional and comfort requirements, facilitating a smoother transition.

Beyond addressing the direct energy needs of molting, maintaining an enriched environment during this transitional phase provides distractions, minimizing stress and subsequent social disturbances within the flock. Incorporating perches, dust bathing areas, and multipurpose foraging opportunities can enrich a hen's daily routine, preventing unnecessary angst.

Through this transformative process, patience and understanding become valuable assets. Witnessing the cycle of loss and renewal heightens appreciation for the resilience inherent in chicken life. By accommodating your flock's temporary needs, you enhance their health and strengthen the bond between keeper and hen.

ENCOURAGING LAYING WITH OPTIMAL COOP CONDITIONS

Creating an optimal coop environment is akin to curating a comfortable sanctuary where hens can lay their eggs freely. Proper nest box placement encourages privacy and tranquility, key to initiating and maintaining productive laying cycles. Position these boxes in serene areas, away from bustling activity to promote a peaceful laying atmosphere. Adequate space, lined with soft bedding, provides comfort and makes eggs less prone to breakage.

Building on this concept, introducing visual diversity in nesting areas—such as varied textures or gentle color contrasts—can create a comforting ambiance for laying hens. Keeping a clean environment while integrating these visual elements can encourage even the most reserved hens to utilize the nesting boxes, ensuring efficient use of all available space.

Consistent cleanliness promotes health and productivity. Regularly maintaining nesting and coop cleanliness deters pests and disease, reflecting care that engenders trust and freedom in hens' laying routines.

Lighting holds sway over egg production, with longer daylight encouraging regular laying. Shorter winter days necessitate supplemental lighting solutions, ideally set on timers to mimic natural patterns. By ensuring 14 to 16 hours of light daily, hens receive cues that spur consistent laying, independent of seasonal changes.

Reducing noise and managing stress are crucial for encouraging egg laying. A quiet and serene coop environment is vital for minimizing distractions and disturbances. Paying attention to the social interactions within your flock and resolving any conflicts can decrease stress levels, creating a harmonious atmosphere that supports consistent egg production.

Engage creatively by integrating sound control practices. Utilizing sound panels or introducing mild soothing background noise, such as recorded chirps or subtle classical music, can foster favorable auditory environments, reducing potential disturbances from external noise sources.

Enhancing coop conditions involves a mosaic of small adjustments that collectively promote a productive environment. From the strategic placement of nest boxes to implementing lighting and ensuring cleanliness, these measures nurture your flock's well-being while optimizing egg production. With each hen comfortably laying, the serene rhythm of chicken life continues amidst a backdrop of care and attention, setting the stage for bountiful harvests day after day.

In closing this exploration into egg production and quality, the stage is set for Chapter 6: Seasonal Care and Adaptation, where the changing seasons bring fresh insights and strategies for maintaining your flock's health and output year-round. With knowledge and practical application, each egg collected is a triumph, encapsulating the harmonious balance achieved between nature and nurture.

CHAPTER 6
SEASONAL CARE AND ADAPTATION

PREPARING YOUR FLOCK FOR WINTER WEATHER

As the first frost nips at your nose, you might be sipping hot cocoa, feeling snug in your wool socks, and pondering how on earth your chickens will fare in the chilly months ahead. Winter presents a unique set of challenges for backyard chicken keepers. But fret not! With a bit of preparation, your feathered friends can enjoy the cold season comfortably. Let's delve into how to keep them cozy and content as the mercury plummets.

Insulation is key to keeping your coop toasty. Think of it as swaddling your chickens' home in a warm, protective blanket. Installing insulation in the coop walls is a great starting point. Recycled materials like foam boards or even old carpets can be incredibly effective, acting as barriers against biting winds. If you're looking for more sustainable options, consider sheep's wool or recycled denim insulation, which offer excellent thermal resistance and are environmentally friendly. For added protection, consider using straw bales around the exterior of your coop. They not only provide excellent insulation but also serve as a windbreak, shielding your flock from those icy gusts that seem to slice right through you.

To enhance the insulation further, think about the coop's roof. A well-insulated roof prevents heat from escaping, and using materials like tar paper can seal in warmth effectively. Additionally, ensuring there are no gaps or leaks in the coop will help maintain a constant and comfortable temperature. Conducting a thorough inspection before the first snowfall can save a lot of hassle later on. Make sure to seal any possible entry points for drafts, as even a small gap can cause significant discomfort for your chickens.

Water management becomes a critical task during winter. The last thing you want is for your chickens' water supply to freeze solid. Heated waterers or electric heaters can be a godsend, ensuring that fresh water is always available. If electricity isn't an option, insulating your water containers with covers can help. A reflective cover with an insulating layer can prevent freezing, keeping the water at a drinkable temperature even when it's frosty outside. Using rubber buckets can also be beneficial as they retain heat better than metal or plastic versions, providing your chickens with an unfrozen drink as they navigate the chilly weather.

Innovative solutions are constantly emerging for keeping water thawed. Consider solar-powered water heaters, which can harness even the feeblest winter sunlight to keep water in a liquid state. Additionally, exploring the option of building thermos-like insulated containers for water storage can thwart the freezing elements. Even adding a few ping pong balls to waterers can prevent surface ice from forming due to their constant motion stirred by the wind.

Dietary modifications for winter are akin to fortifying your chickens against the cold. Increasing their calorie intake with cracked corn provides the extra energy they need to stay warm. Think of it as giving them an extra slice of pie at Thanksgiving. Additionally, foods high in fat and protein, such as sunflower seeds or kitchen scraps like oatmeal, can be introduced to give them an energy boost. Warm mash meals also serve as a treat, especially on those bone-chilling mornings. A mix of feed with warm water not only warms their bellies but also encourages them to eat more.

In the winter, when the ground is frozen and fresh greens are hard to come by, bringing the garden indoors can make a significant difference in your chickens' diet and health. Implementing a hydroponic system allows you to grow nutrient-rich microgreens, such as alfalfa, clover, and radish, right inside your home or a suitable outbuilding. This method of growing fodder bypasses the need for soil, using water enriched with minerals instead, and can yield a fresh supply of greens in just a matter of days. Including these microgreens in your chickens' diet not only introduces essential vitamins and minerals that may be missing during the colder months but also provides them with a varied diet, promoting better health and egg production.

Providing proper shelter and bedding ensures your chickens remain snug and dry. The deep litter method is an excellent way to generate warmth within the coop. By allowing bedding to accumulate and decompose, natural heat is produced, similar to a compost pile. Regularly adding fresh layers of straw or wood shavings will keep things clean and cozy. Ensuring dry bedding is crucial to prevent frostbite on those delicate chicken toes. Dampness is the enemy, so check often and replace bedding as needed. Consider using pine needles or shredded leaves for added insulation and a pleasant scent that can double up as natural mite repellents.

To supplement the deep litter method, explore using heat lamps cautiously. While they can provide targeted warmth, they also pose a fire risk if not positioned correctly. Consider using ceramic heat emitters that provide warmth without light, reducing stress while ensuring safety. Always integrate these systems with thermostat controls to maintain optimal temperatures.

WINTER READY CHECKLIST

- **Insulation**: Ensure coop walls are insulated; add straw bales or consider alternatives like sheep wool for wind protection.
- **Water Management**: Use heated waterers or insulated covers for water containers.
- **Diet**: Increase calories with cracked corn; offer warm mash, sunflower seeds, or oatmeal meals.
- **Shelter**: Implement the deep litter method; maintain dry bedding with additional layers like pine needles for warmth.

This checklist serves as a handy guide to keep your flock warm and comfortable despite the cold months. By addressing these aspects of winter care, you'll provide a secure, inviting environment where your chickens can thrive, despite the cold. And while winter might bring challenges, it also brings opportunities to deepen your connection with your flock as you work together to navigate the chill.

KEEPING CHICKENS COOL DURING HOT SUMMER MONTHS

Imagine a sun-drenched afternoon, the air thick with heat, and your flock looking slightly less perky than usual. Chickens, much like us, don't appreciate sweltering days. The good news is you can create a refreshing oasis for them with a few thoughtful adjustments. Start with promoting shade and ventilation. Installing fans or misters in the coop can work wonders to keep the air circulating. These aren't just luxuries—they're lifesavers when temperatures soar. If installing technology feels daunting, creating shaded areas with simple tarps or strategically planted trees can offer relief from the sun's relentless rays. Your chickens will thank you with every contented cluck.

Beyond artificial shade options, explore planting fast-growing trees that lose their leaves in winter but provide ample shade in summer, such as willows or poplars, strategically placed around the coop. This not only offers year-round shade benefits but will also enhance the coop's landscape aesthetics.

Hydration is another cornerstone of summer chicken care. Water isn't just about quenching thirst; it's essential for regulating body temperature. To ensure your chickens stay hydrated and healthy, especially during the warmer months, consider incorporating electrolytes into their water supply. Electrolytes are vital for maintaining a chicken's fluid balance and can help replenish the essential minerals they lose through exertion in the heat. This simple addition to their water can significantly impact their overall well-being. It's akin to sipping an ice-cold sports drink after a sweaty jog.

Frozen treats like watermelon slices can also provide a refreshing break and add a bit of excitement to their day. Picture your flock pecking eagerly at chilled fruity goodness—a delightful scene that keeps them hydrated and happy. Ice blocks with embedded fruits or vegetables can provide an engaging way for your flock to cool off while having some nutritious fun.

An alternative hydration method involves the use of nipple waterers or automated drinkers that ensure a constant supply of fresh, cool water. Consider installing a water filtration system to provide the purest refreshment, minimizing the risk of bacteria which often proliferate in heat.

Diet is another area where small tweaks can make a big difference. During summer, reducing corn intake is wise, as corn generates internal heat during digestion. Opt instead for high-water-content foods. Cucumbers, lettuce, and zucchini are excellent choices that provide hydration while being easy on their digestion. Think of it as swapping out hot soup for a crisp salad on a summer day—it just makes sense. Cold grains like barley can also be introduced, offering a nutritious and cooling alternative to the usual mix, ensuring that your chickens have the energy necessary without the added heat.

Moreover, the introduction of live insects or high-protein snacks ensures sustained energy levels for active summer days. Engaging your chickens with a diet rich in variety can further reduce stress and encourage healthy behavior.

Observing your chickens closely is key to catching signs of heat stress early. Panting, lethargy, and holding wings away from the body are chicken ways of telling you it's too darn hot. They might even look like they're practicing yoga with their wings stretched out like that! Recognizing these signs lets you step in before any real harm occurs. Strategically placed shade cloths or low-lying plants can also encourage your flock to find the cooler parts of the yard to retreat to in times of extreme heat.

Integrating heat-tolerant foraging plants or grasses into your run area can provide natural retreats and even nutritional benefits. Purslane and other drought-resistant plants can be grown to offer a cool respite and a source of omega-3-enriched greens.

SUMMER HEAT CHECKLIST

- **Shade & Ventilation**: Install fans/misters; create shaded spots with tarps, trees, or vegetation.
- **Hydration**: Add electrolytes to water; provide frozen treats like watermelon or ice blocks with embedded goodies.
- **Diet Adjustments**: Reduce corn; offer high-water-content foods and consider cold grains like barley.

- **Behavior Observation**: Look for panting, lethargy, or wings held away from the body, and encourage rest in cooler areas.

This checklist serves as a quick reference to keep your chickens comfortable and healthy throughout those sizzling summer months. The beauty of these strategies is that they're simple yet effective, allowing you to enjoy summer knowing your flock is well cared for. By addressing the needs of your chickens before the heat intensifies, you ensure that they remain cool, calm, and clucking contentedly all through the season.

ADJUSTING CARE ROUTINES FOR SPRING AND FALL

As the first green shoots of spring poke through the soil, it's time to adjust the care routines for your flock. Spring and fall are transitional seasons, each with their quirks and challenges. Both demand flexibility and awareness. One of the first tasks is gradually adjusting light exposure. Chickens are sensitive to daylight changes, which can affect their laying patterns and behavior. As days lengthen in spring, you might notice a boost in egg production, with hens busily scurrying to settle in their nests. In contrast, fall brings shorter days, signaling your hens to slow down. To help with this transition, gradually increase or decrease artificial lighting in the coop, mimicking natural daylight changes.

Fine-tuning artificial light in the coop not only aids in egg production but helps provide a consistent sleep cycle which can reduce stress. Opt for a timer-based lighting system to emulate natural crepuscular light gradients which can have calming effects on your flock.

Weather can be unpredictable during these seasons, with temperatures swinging wildly from warm to chilly in a matter of days. Monitoring these sudden shifts is crucial. Your chickens can get stressed from abrupt changes, affecting their health and productivity. Keep an eye on weather forecasts and adjust coop ventilation accordingly. On warmer days, ensure proper airflow to keep your flock comfortable. During chilly spells, close vents to retain warmth without sacrificing ventilation. Keeping the coop properly ventilated helps prevent respiratory issues that can arise from poor air quality.

Ensure your chicken coop is equipped with adaptive ventilation systems that respond to changes in humidity. Consider installing automatic vent flaps or solar-powered windows, which open and close based on temperature fluctuations, to effortlessly sustain the ideal conditions for your chickens.

Spring and fall also herald the molting season—a time when chickens shed old feathers and grow new ones. This process is energy-intensive, and your flock will need extra protein to support feather regrowth. Consider supplementing their diet with high-protein snacks like mealworms or sunflower seeds. These tasty treats will help them through this demanding phase. During molting, chickens become more vulnerable and may seek out safe spaces away from the flock. Ensure there are quiet corners in the coop where they can retreat without being disturbed by more boisterous companions. Providing extra bedding and low-light areas can help them feel secure during this somewhat uncomfortable period.

Supplement molting diets with probiotics and vitamins, ensuring the new growth is robust and healthy. Linseed and flaxseed can offer omega-3s for improved feather quality, supporting radiant plumage post-molt.

Parasite prevention becomes especially important during these transitional seasons. As temperatures change, parasites become more active, seeking out hosts in your flock. Regular coop cleaning can disrupt their life cycles, making your chickens less attractive targets. Scrub down surfaces and change bedding frequently to reduce parasite load. Natural deterrents like diatomaceous earth can also help. Sprinkle it in nesting boxes and along coop borders as a first line of defense against mites and lice. Keeping herbs such as lavender or peppermint near the coop can act as natural repellents, enhancing both scent and health within your chickens' living quarters.

Incorporate natural insect interceptors, like predator nematodes, into the soil around your coop to target larvae before they reach adulthood. Encouraging beneficial insects aids in keeping the parasite population under control naturally.

Springtime often brings a surge in egg production, and you'll need to prepare for this bounty. Consider adding extra nesting boxes to accommodate the increased output. This prevents overcrowding in existing boxes, reducing stress for your hens. Enhanced nutrition is another key factor during this period. Laying hens require additional calcium and nutrients to produce strong eggshells consistently. Layer feed supplemented with oyster shells or crushed eggshells provides the necessary boost. Checking water schedules is essential during increased egg production, as more eggs mean more water needed for your flock's sustenance and overall health.

Utilize crushed limestone as a regular supplement for calcium to aid in egg production. Introducing crushed eggshells back into your chickens' diet closes the cycle of use and replenishment sustainably.

Balancing these seasonal changes requires an eye for detail and a bit of patience. But with thoughtful preparation, you can ensure your flock thrives through spring and fall's ups and downs. Your chickens will reward you with robust health and plentiful eggs as they adapt to the rhythms of nature's cycles. As you observe their behaviors and needs, you'll find that these transitional periods are opportunities for growth—for both you and your flock.

The beauty of chicken-keeping lies in its ever-changing nature, with each season bringing new challenges and joys. As you tweak routines and make adjustments, remember the satisfaction that comes from caring for your feathered friends. Whether it's watching them scratch happily in the first warm days of spring or seeing them settle contentedly into cozy nests as autumn leaves fall, each moment spent nurturing your flock deepens your connection to this rewarding endeavor.

PROTECTING CHICKENS FROM PREDATORS YEAR-ROUND

I know we've touched on predator awareness in earlier chapters, but its importance cannot be overstated, so it bears addressing again here. Specific threats can vary throughout the year, and the change in seasons offers an opportune time to reassess our security measures.

The arrival of spring is often heralded by the cheerful chirping of newborn chicks and the gentle rustling of leaves. Yet, it also brings a different crowd—foxes on the prowl. As spring unfolds, foxes become more active, driven by the need to feed their young. This uptick in activity means your chickens might be on their radar. Similarly, in the fall, migrating hawks take to the skies, posing another threat as they hunt for easy prey during their long journeys. Understanding these seasonal shifts in predator behavior is key to safeguarding your flock. You want to stay one step ahead, knowing when to tighten security around your coop.

Consider scheduling routine patrols during high-risk periods, especially at dawn and dusk when predators are most active. This vigilance can deter potential intrusions before they occur.

Reinforcing your chicken coop's defenses doesn't have to be a Herculean task. Think of it as giving your fortress a yearly tune-up. Check fences for wear and tear, ensuring they're strong enough to withstand a determined predator's attempts to breach them. Repair any gaps or weak spots promptly. Installing secure locks on coop doors is another must-do. Raccoons are notorious for their dexterity, and a simple latch won't deter them. Opt for a lock that requires a bit of human ingenuity—a combination lock or a carabiner clip can work wonders.

Install predator-proof wiring that extends several inches into the ground around the coop. This barrier technique helps prevent digging predators, such as foxes, from tunneling their way in.

Sometimes, deterrence is the best form of defense. Motion-activated lights or alarms can startle would-be predators, sending them scurrying back into the shadows. The sudden burst of light or noise can convince them that your coop isn't worth the trouble. For aerial threats like hawks, consider adding predator-proof netting over your run. It's like putting up an invisible shield that keeps your chickens safe from above. This netting is sturdy yet unobtrusive, allowing sunlight to filter through while keeping danger out. Reflective metallic tape can also be strategically placed to deter predators with its startling, shifting light patterns.

Embrace technology by installing smart home cameras around the coop. These can alert you directly through your phone of any suspicious activity, allowing for rapid response and peace of mind irrespective of where you are.

Creating a community of fellow chicken keepers can be an invaluable resource in your quest to protect your flock. Networking with local chicken enthusiasts not only provides support but also creates a shared defense network. Keeping each other informed about recent predator sightings in your area can be a game-changer. You might hear about a neighbor who spotted a fox near their coop or learn that a hawk has been circling overhead. This information allows everyone to stay vigilant and adjust their strategies accordingly. Hosting a quarterly meet-up to share experiences and tips can provide fresh insights and foster a supportive environment for addressing predator issues collectively.

Community initiatives such as neighborhood watch groups or shared patrolling schedules further solidify a defense strategy, spreading knowledge and offering protection through collective effort.

PREDATOR DEFENSE CHECKLIST

- **Seasonal Awareness**: Stay alert to fox activity in spring and hawk migrations in fall.
- **Barrier Enhancements**: Regularly repair fences; install secure locks on coop doors.
- **Deterrents**: Use motion-activated lights/alarms; install predator-proof netting and reflective deterrents.
- **Community Support**: Network with local keepers; share predator sightings and tips.

This checklist serves as a quick reference to bolster your flock's security year-round. Through mindful preparation and community collaboration, you can create a formidable defense against the ever-present threat of predators. Recognizing the patterns of nature and adjusting your approach accordingly ensures that your chickens remain safe, allowing you to continue enjoying the many rewards of backyard chicken-keeping. With these strategies in place, you're not just reacting to threats—you're proactively creating an environment where your flock can thrive in peace.

Interwoven with the rhythms of the seasons are challenges that demand vigilance and creativity. Each step you take to protect your chickens reflects your commitment to their safety and well-being. The satisfaction of watching them scratch and peck without fear is a testament to the efforts you've invested in their care. In this dance with nature, your role as protector is ever-evolving, adapting to the changes with resilience and resourcefulness.

ADAPTING COOPS FOR SEASONAL CHANGES

As the seasons change, so does the environment around us. Your chickens feel it too, and adapting your coop to meet these shifts can make a world of difference. Imagine your coop as a living space that needs a wardrobe change with each season. Removable panels are a brilliant solution for ventilation control. During the summer, these panels can be removed to allow fresh air to flow through, helping to keep the interior cool. In winter, simply snap them back into place to keep the chill out. This flexibility means you're always ready for whatever Mother Nature throws your way.

Explore using magnetic or easy-slide panels for swift seasonal adjustments. These innovative designs cut down on time and effort while maintaining efficient changeover.

Seasonal roofing adjustments are another way to keep your chickens comfortable year-round. In the rainy months, consider adding a sloped roof extension or even temporary awnings to direct water away from the coop. This prevents leaks and keeps things dry inside. When the sun's blazing in summer, a reflective roof coating can help deflect heat, keeping the coop cooler. These tweaks are like giving your coop a seasonal makeover, ensuring it's always prepared for the weather ahead.

Additionally, consider installing retractable roofs or vents that can adjust based on moisture sensors, optimizing ventilation during downpours or heatwaves automatically.

Managing the coop climate effectively is all about balance. During the colder months, using heaters can provide much-needed warmth, but it's crucial to ensure they are safe and don't pose a fire risk. Conversely, fans can circulate air during warmer times, preventing overheating and reducing humidity buildup. Adjusting insulation is also essential for moderating temperatures. You might find that adding or removing layers of insulation as seasons change helps maintain a steady internal environment, reducing stress on your chickens.

Integrate solar-powered climate control solutions that adjust heating and cooling systems automatically. These self-sufficient technologies not only enhance comfort but reduce reliance on the grid, promoting an eco-friendly setup.

Routine maintenance is your secret weapon in keeping the coop running smoothly through all seasons. Regular inspections can catch small issues before they become big problems. As autumn leaves fall and spring showers hit, check for leaks or drafts that may have appeared. Cleaning gutters and ensuring drainage systems are clear will keep water from pooling around the coop—nobody likes soggy feet, least of all your chickens! These simple checks and repairs are like giving your coop a seasonal health check-up.

Schedule monthly cleanup sessions that involve thorough cleaning and checks of moving parts or flexible components, ensuring that all mechanisms operate smoothly without lag or obstruction.

Creative design solutions can make these seasonal adjustments easier and more effective. Convertible coop designs offer multi-seasonal functionality with minimal effort. Imagine a coop with sections that can be easily reconfigured to suit different weather conditions. It's like having a Swiss Army knife of coops—versatile and ready for anything! Whether it's adding a removable sunshade in summer or swapping out heavier blankets for lighter ones as spring arrives, these innovations make managing your flock easier.

Engage in modular design experiments that allow for customized coop setups, meeting diverse needs without over-hauling existing infrastructure. Tailored solutions ensure an ideal setup every season.

Inspiration can come from anywhere when it comes to coop design. Perhaps you've seen a neighbor's clever setup or stumbled across an innovative idea online. Don't be afraid to experiment and make your coop uniquely yours. Adding solar panels for lighting or rain barrels for water collection are great ways to enhance sustainability while catering to your flock's needs.

Remember that your chickens rely on you to create an environment where they can thrive regardless of the season. By making thoughtful adaptations and staying proactive with maintenance, you ensure their comfort and safety year-round. Plus, you'll find immense satisfaction in knowing you've created a well-oiled machine that serves both you and your feathered friends beautifully.

As you tweak and tinker with your coop, each change brings new learning experiences and opportunities for creativity. The beauty of adapting your setup lies in its ability to reflect not just the changing seasons but also your evolving understanding as a chicken keeper. Every adjustment is an act of care and commitment to your flock's well-being, showcasing the deep bond between you and these delightful creatures who share your space and life.

MANAGING SEASONAL BEHAVIOR CHANGES IN CHICKENS

As the days grow shorter and the sun dips below the horizon earlier, you might notice your chickens acting a bit differently. It's almost like they're signaling a shift in mood, much like how we feel a little off when summer turns to fall. One of the most noticeable changes is increased aggression during those shorter daylight hours. Your usually docile hens might become more territorial, vying for the best roosting spots or first dibs on feed. This behavior is often linked to the reduced light affecting their natural rhythms.

Individual tracking of aggressive behavior enables targeted interventions. Document specific cases with details on frequency and triggers, ensuring a tailored approach that minimizes disruptions.

Environmental stresses, such as sudden cold snaps or heatwaves, can also shake up the pecking order within your flock. When chickens feel under pressure, they often resort to reestablishing their social hierarchy as a way to cope. You may find that a previously dominant hen is suddenly challenged by a younger upstart. These shifts can cause temporary unrest, but with time and patience, things usually settle down.

To help manage your chickens' stress during periods of change or upheaval, consider implementing calming procedures that can significantly ease their temperaments. One effective method is to create a dimly lit environment, which can have a soothing effect on your flock. Additionally, the use of chicken-safe essential oils, such as lavender or chamomile, diluted in water and lightly misted in the coop, can also promote a sense of calm among your birds. Always ensure that any essential oils used are safe for chickens and applied in a manner that does not overwhelm them.

To help your flock navigate these changes, implementing stress-reduction techniques can be incredibly effective. Maintaining consistent routines and feeding schedules provides your chickens with a sense of stability. Think of it as sticking to a familiar routine when everything else seems unpredictable. Enrichment activities can also work wonders in keeping stress levels down. Scatter some scratch grains around the run or hang a cabbage for them to peck at —it's like giving them a puzzle to solve.

Incorporating interactive foraging systems into your chicken coop introduces a range of rotating challenges that are vital for both the mental and physical stimulation of your chickens. These systems, which feature various obstacles and feeding puzzles, are designed to replicate the natural foraging behaviors of chickens in the wild. They encourage your chickens to engage in activities like exploring, pecking, and scratching, mirroring their instinctual habits. This engagement is not only beneficial for keeping your chickens active but also plays a significant role in preventing bore-dom-induced conflicts within the flock. Implementing these enrich-ment strategies can be key to fostering a peaceful and healthy environment in your chicken coop.

During these times, adapting social management practices becomes crucial. If you notice one bird being particularly aggres-sive, it might be time to give her a little time-out. Separating aggressive birds temporarily can allow tensions to cool down without escalating into full-blown disputes. In the meantime, use distractions to reduce tension. A few strategically placed mirrors or shiny objects in the coop can captivate their attention and redirect energy away from pecking at each other.

For persistent aggression, explore integration training, focusing on socialization with unfamiliar environments or animals to reverse undesirable behaviors gradually.

Monitoring health and activity levels is equally important in managing seasonal behavior changes. Regular health checks help you catch any seasonal issues early. Look for signs of illness or injury that might be exacerbated by stress or weather changes—a droopy comb or a listless demeanor could indicate a problem. Keeping a record of behavioral observations allows you to spot patterns over time. It's like being a detective piecing together clues about your flock's well-being.

BEHAVIOR OBSERVATION JOURNAL

Document daily interactions and note any changes in behavior or social structure. Record incidents of aggression, health concerns, and responses to stress-reduction strategies.

This journal not only serves as a record but also helps you tailor your care approach based on your chickens' unique needs.

In this ever-changing dance with nature, understanding and responding to your chickens' behavioral cues fosters a harmonious environment. By recognizing shifts, reducing stress, and monitoring health, you are not only ensuring your flock remains healthy and content throughout the seasons but also gaining deeper insight into their unique personalities. You'll find that each chicken is an individual, as distinct as the changing seasons themselves, with their habits and quirks.

As we wrap up this chapter, remember that each season brings its own set of challenges and opportunities for growth—both for your chickens and for you as their caregiver. Embrace these changes with flexibility and patience, knowing that each adjustment strengthens your bond with your flock. In the next chapter, we'll explore how to seamlessly integrate your chickens into a sustainable lifestyle that benefits both them and the environment. As a guardian of your flock, your adaptability not only aids in their survival but enhances the quality of life for everyone under your care, enriching your understanding of nature's intricate and beautiful cycles.

CHAPTER 7
INTEGRATING CHICKENS INTO SUSTAINABLE LIVING

BENEFITS OF FREE-RANGE CHICKENS FOR SUSTAINABLE LIVING

Envision a serene morning, enjoying the fresh air while watching your chickens roam and explore the backyard freely. Their feathers catch the light as they cluck and scratch, pecking at insects and seeds hidden in the grass. This is the essence of free-ranging, where chickens exhibit natural behaviors that promote their health and happiness. Allowing chickens to forage not only enriches their diet but also keeps them engaged and active. As they hunt for bugs and seeds, they exercise naturally, boosting their overall well-being. Let's not forget dust bathing—watching a chicken kick up dust as they roll around is a joy. This natural behavior keeps their feathers clean and their skin free from parasites.

Imagine the long-term effects of these natural habits. By engaging in natural foraging, chickens develop stronger immune systems, making them less susceptible to diseases, which consequently reduces the necessity for antibiotics or other medications.

Their robust health can lower potential veterinary costs and contribute positively to the overall ecosystem of your backyard. With each scratch and peck, they turn the soil, facilitating aeration, which over time, improves the garden's micro-environment.

Beyond their personal happiness, free-ranging chickens offer significant environmental benefits. By allowing chickens to roam, you reduce the need for manufactured feed, which cuts down on the resources needed to produce and transport commercial feed products. As they wander, they naturally fertilize the ground with their droppings, enriching the soil without synthetic fertilizers. This contributes to a more sustainable cycle, where waste becomes a resource rather than a problem. The chickens' droppings also improve soil structure, which can increase water retention and reduce erosion. This natural fertilization decreases the overall waste footprint of your backyard operation.

Consider the broader impacts on your community—when more people adopt free-range practices, the collective reduction in feed production and transportation could significantly lower carbon emissions. In essence, your choice to free-range chickens can ripple outwards, influencing broader environmental changes and fostering a culture of sustainability.

But the benefits don't stop at environmental impact; there's a nutritional advantage too. Eggs produced by free-range chickens often boast higher levels of omega-3 fatty acids, which are vital for heart health. The varied diet that free-ranging affords leads to eggs with richer yolk colors, indicating a better nutrient profile. These eggs not only taste better but also offer more nutritional value compared to those from confined hens. The meat from chickens raised on pasture can also be leaner and richer in flavor, making it a healthier choice for your table.

Picture a Sunday breakfast with the deep golden yolks of free-range eggs glistening on your plate. The satisfaction of knowing that these come not only from healthy chickens but also contribute positively to your health is unmatched. This fulfillment enhances your connection to your food sources, promoting a more conscientious and satisfying approach to eating.

Real-world examples abound in small-scale free-range operations showing increased egg production and improved chicken health. Consider a local farmer who switched from conventional to free-range and saw a boost in egg quality and output. Their chickens became more robust and lively, reflecting the positive impact of a natural diet and environment. Testimonials from other chicken keepers highlight similar stories: healthier flocks, tastier eggs, and a deeper connection to sustainable living practices. These successes illustrate how free-ranging can transform not just the quality of produce but also the joy of chicken keeping itself.

Engage with these examples and consider joining networks of free-range growers. Sharing experiences, successes, and challenges within these communities can provide support, inspire innovation, and reinforce your journey toward sustainable chicken-keeping.

Connect with local agricultural extension services or attend workshops to deepen your understanding of free-range techniques. These opportunities can broaden your perspective and introduce you to like-minded individuals dedicated to sustainable agriculture. Engaging with organizations that promote sustainable agriculture can provide a wealth of information and resources, further supporting your free-range endeavors.

USING CHICKENS IN PERMACULTURE AND GARDEN ECOSYSTEMS

The principles of permaculture revolve around creating sustainable systems that mimic natural ecosystems, and chickens fit perfectly into this vision. They act as natural pest control agents, eagerly seeking out insects that might otherwise plague your plants. By allowing chickens to roam among your garden beds, you enlist their help in managing pests without the need for chemical solutions. As they scratch and peck, they naturally aerate the soil, improving its structure and promoting nutrient flow. This symbiotic relationship enhances both the garden's vitality and the chickens' access to a varied diet.

Envision extending this dynamic partnership by incorporating chickens into your crop rotation plans. They can be moved into a plot after harvest where they help prepare the soil for the next planting season by eating leftover plant material and pests, turning your whole garden into a linked, self-sustaining system.

Incorporating chickens into garden spaces requires a thoughtful approach to protect your crops. Rotational grazing is a technique that prevents over-foraging by moving chickens between different garden sections. This method allows vegetation to recover while spreading the benefits of chicken activities throughout your plot. Portable fencing serves as an invaluable tool, granting you precise control over where your flock wanders. By redirecting their foraging paths, you ensure that your plants remain intact and thriving.

Strategically planting certain crops can create a mutually beneficial environment for both chickens and gardens. Comfrey, with its deep roots and high nutrient content, is an excellent choice. Chickens love to browse its leaves, and when it dies back, it adds valuable nutrients to the soil. Similarly, herbs like basil and mint repel pests that might otherwise damage your plants. By placing these aromatic herbs near vulnerable crops, you create a natural barrier against common garden pests.

Visualize the design of your garden layout with these considerations in mind—your garden could be a thriving ecosystem where every element supports the health of others. Sketch a map of your space to plan rotations and strategically place pest-repelling plants.

Imagine a garden where every element works together in harmony. Diagrams of integrated systems can help visualize how to achieve this balance. Picture a permaculture setup with raised beds for vegetables surrounded by pathways for chicken access. Portable fencing might section off areas for specific plantings, allowing chickens to roam freely without damaging delicate seedlings. Photos of thriving gardens can inspire you to design a space that suits your unique landscape while maximizing the benefits of chicken integration.

Creating a balanced ecosystem with chickens involves more than just letting them wander freely. Managing flock size ensures that your plant resources match chicken needs. Overgrazing can strip an area bare, so rotating their grazing areas helps maintain lush growth. Introducing biodiversity into your garden enhances resilience against pests and diseases.

Incorporate plant species native to your area to enhance your chicken-friendly garden ecosystem. These indigenous plants not only demand less maintenance but also bring significant environmental advantages. By choosing local flora, you create a natural habitat that supports a diverse population of beneficial insects and pollinators. These inhabitants play a crucial role in maintaining a balanced ecosystem by keeping pest populations under control, thus reducing the need for chemical pesticides. Furthermore, native plants are well-adapted to your garden's soil, moisture, and climate conditions, making them more resistant to diseases and pests. This resilience contributes to a healthier, more sustainable garden environment that supports your chickens in their foraging activities. They benefit from a richer and more diverse diet by having access to the wide variety of insects and plants that these native species attract, enhancing the nutritional value of their eggs and meat. Integrating indigenous plants into your garden not only supports your chickens' health and well-being but also promotes biodiversity and strengthens the local ecosystem.

Enhancing the habitat with shrubs and grasses offers chickens shelter and shade while adding visual interest to your garden.

Shrubs like elderberry or currant provide cover from predators and extreme weather while offering fruits for both you and your flock to enjoy. Grasses contribute grazing opportunities, creating a rich tapestry of textures and colors. Regular soil testing guides adjustments in planting strategies, ensuring optimal fertility year-round.

Diversifying your approach can significantly improve the efficiency and output of your chicken-keeping activities. Designating a specific section of your garden to create a tree guild represents a thoughtful strategy for fostering a balanced ecosystem.

A tree guild, at its core, is designed around a central tree chosen for its beneficial attributes, such as fruit or nut production, which serves as the anchor for the surrounding plant community. Around this key tree, companion plants are carefully selected and arranged in a way that they support each other's growth while maximizing the use of space. This could include smaller fruit or nut trees, berry bushes, beneficial herbs, and groundcover plants, each contributing to the guild in unique ways—some fix nitrogen in the soil, others attract beneficial insects, or provide mulch through leaf drop. Together, they create a symbiotic environment that requires less maintenance over time, as the plants' interdependent relationships reduce the need for fertilizers and pest control.

Incorporating a tree guild into your chicken-friendly garden not only enhances the biodiversity and resilience of your garden ecosystem but also offers an enriched foraging area for your chickens, promoting their health and well-being. This structure promotes a symbiotic relationship among plants, providing benefits like shade, shelter, and nutrients. By doing so, you make the most of limited space while fostering a supportive environment for your chickens and garden alike.

To involve your family in these activities, organize weekend projects like building a simple portable chicken coop together, incorporating fun challenges or games to make sustainability an engaging family affair.

Composting chicken manure is another cornerstone of integrating chickens into a sustainable ecosystem. This process transforms waste into valuable fertilizer by balancing nitrogen-rich manure with carbon materials like straw or leaves. Aerating compost piles accelerates decomposition, providing nutrient-rich soil amendments that enhance garden fertility. Chicken manure's high nitrogen content fuels leafy growth while improving soil structure and water retention.

Reducing waste through efficient practices further aligns chicken-keeping with sustainable living goals. Recycling chicken bedding as mulch or compost enriches soil while minimizing landfill contributions. Kitchen scraps become valuable feed, reducing food waste while supplementing poultry diets. Creative repurposing extends even to feathers, which can be crafted into insulation or art projects.

Innovative recycling methods highlight the potential for resourcefulness in chicken-keeping. Feathers can be transformed into crafts or used as insulation material. Chicken litter can undergo pyrolysis to create biochar, adding long-term carbon storage to the soil while enhancing fertility. These techniques exemplify how embracing sustainability fosters creativity and reduces environmental impact.

Engaging with local communities amplifies the benefits of sustainable chicken-keeping. Organizing coop tours fosters connections among fellow enthusiasts, sharing knowledge and inspiration. Workshops on sustainable practices empower others to adopt eco-friendly methods in their own backyards. Online platforms provide spaces for exchanging ideas and troubleshooting challenges together.

Collaborative projects strengthen community bonds while promoting resource sharing. Co-op purchasing reduces costs for feed and supplies while supporting local economies. Collective composting initiatives distribute resources efficiently, turning potential waste into nourishment for gardens across neighborhoods.

Success stories from communities thriving through shared efforts inspire others to join the movement toward sustainable living with chickens at its heart—whether through community gardens incorporating poultry or urban initiatives advocating for backyard flocks as part of resilient cityscapes.

Through integrating chickens into permaculture designs and community efforts alike, you contribute positively not only to your immediate surroundings but also toward broader ecological goals —transforming simple backyard endeavors into impactful steps toward sustainability on both local and global scales.

Leverage social media or local events to create a buzz around sustainable chicken-keeping practices, inviting people to explore and adapt these strategies to their contexts.

CREATING A BALANCED ECOSYSTEM WITH CHICKENS AND PLANTS

Watching chickens roam the yard, it struck me how much their presence transformed the landscape. They became part of a delicate balance, a dance between feathered foragers and the greens they care for so unintentionally. Maintaining this harmony is crucial. Managing flock size is a start; too many chickens can strip an area bare, leaving the ground barren and depleted. It's vital to match the number of chickens to the resources available, ensuring that plants have time to recover and thrive. Rotating chicken areas keeps them from overgrazing a single spot and allows vegetation to replenish. Think of it like crop rotation but with your flock, keeping both plants and chickens in prime condition.

Diversity should be embraced not just in plant and flower species but also in landscaping elements like ponds or rock gardens. These features attract helpful creatures like dragonflies or frogs, which further contribute to pest control, adding layers to your ecosystem's balance.

An ecosystem thrives on diversity, and introducing a variety of plant and animal life creates resilience. Native plants play a significant role in this balance, offering habitat support for local wildlife and beneficial insects. These insects, in turn, help control pests that might otherwise target your garden. By incorporating flowers that attract pollinators, you invite nature's helpers to your yard. The symbiotic relationship between chickens, plants, and insects forms a web of interactions that support ecosystem health. You'll find that a diverse garden isn't just more beautiful but also more robust against diseases and weather changes.

A wide variety of plants ensures that even if one species struggles, others can maintain the integrity of the ecosystem, proving the importance of biodiversity in resilient gardening and chicken-keeping.

Enhancing chicken habitats with plantings adds both beauty and function to your space. Consider shrubs like elderberry or lilac for shelter and shade. These not only provide your chickens with protection from the sun and predators but also bring vibrant color to your garden. Grasses such as rye or fescue offer grazing opportunities and ground cover. They're easy on the eyes and gentle on the soil, preventing erosion while feeding your flock. Creating these micro-habitats encourages chickens to explore and engage with their environment, making for happier birds.

Practical tips go a long way in keeping your ecosystem balanced. Regular soil testing can tell you a lot about fertility levels and whether your plants are getting the nutrients they need. It's a simple process but incredibly informative, allowing you to make informed decisions about what to plant where. Adjusting plantings based on seasonal changes is another key strategy. Some plants thrive in the cool of spring or fall, while others bask in summer's heat. By rotating crops and adjusting your chicken's access accordingly, you maintain soil health and plant vitality.

Consider setting up marker layouts or logs detailing which chickens access which sections. Over time, these records become indispensable in your decision-making, streamlining processes and informing strategies for upcoming seasons.

PRACTICAL TIP: SOIL TESTING FOR FERTILITY

To integrate soil testing into your garden management, procure a soil test kit from a nearby garden center or cooperative extension office. Follow the enclosed instructions meticulously to collect soil samples from different areas of your garden. The analysis will illuminate the nutrient content, pH levels, and any required soil amendments to enhance plant growth.

Balancing an ecosystem with chickens requires attentiveness and adaptability. It's about observing how each element interacts with others and making adjustments as needed. You might find that one area of your yard becomes a favorite dust bathing spot or that certain plants become targets for enthusiastic pecking. These observations guide your actions, helping you create an environment where all elements coexist harmoniously.

In this living tapestry, chickens play a vital role in maintaining ecological balance. Their natural behaviors contribute to soil health, pest control, and nutrient cycling. By understanding their impact and managing it thoughtfully, you foster an ecosystem that supports both plant and animal life. It's an ongoing process—a dynamic interaction that evolves with time and experience.

Observe how various areas of your garden weather different challenges. You may notice one section recovers quicker after heavy rain, indicating a potential for increasing rainwater harvesting opportunities in your chicken-keeping strategy.

With each season, you'll learn more about what works best in your space. Some years may bring surprises—unexpected pests or unusual weather patterns—but these challenges are opportunities for growth. By embracing diversity, enhancing habitats, and staying attuned to the needs of your garden and flock, you build a resilient system that thrives year after year. By reinforcing a connection between the garden and its chicken inhabitants, you form a more holistic approach to permaculture.

This approach turns chicken keeping into more than just a hobby; it becomes part of a larger commitment to sustainable living. It's about creating a space where chickens aren't just residents but integral members of a vibrant ecosystem. Whether you're new to this or have been at it for years, there's always more to discover and enjoy in this rewarding pursuit.

COMPOSTING WITH CHICKEN MANURE FOR GARDEN FERTILITY

Composting chicken manure is like turning a waste product into gardening gold. It's a straightforward process, but understanding the basics is crucial. Chicken manure is rich in nitrogen, which is fantastic for leafy growth, but it can be too potent if used directly on plants. The trick is balancing this nitrogen-rich manure with carbon materials like straw or dried leaves. Think of it as creating a perfect recipe, where ingredients must be carefully proportioned to achieve the best result. The carbon materials help to mellow out the strong effects of the nitrogen, making sure your plants get the nutrients without the burn. Aerating your compost pile is another key step in decomposition. It's much like fluffing a pillow—it keeps things light and allows oxygen to circulate, which helps microorganisms break down the material faster.

Involving family members in these processes can make composting a community effort, spreading enthusiasm and knowledge about the benefits of natural fertilizers among all age groups. Creating compost bins can be a family-friendly project where even youngsters learn the crucial part they play.

Chicken manure is a powerhouse for enriching garden soil. Its high nitrogen content fuels lush, green growth, making it ideal for leafy vegetables and other plants craving nitrogen. But that's not all. It also enhances soil structure by improving aeration and water retention. This means your garden will hold moisture better, reducing the need for frequent watering and increasing resilience during dry spells. Plus, composted chicken manure introduces beneficial microorganisms to the soil, boosting its overall health and fertility. These microbes are like little helpers working tirelessly to keep your garden thriving.

Further optimize your soil management by rotating the types of crops you grow, allowing your chicken manure compost to support varying nutrient demands while maintaining soil vitality year-round.

There are different methods to manage your manure, each offering unique benefits. Hot composting is one technique that speeds up the decomposition process. By maintaining a tempera-ture between 140°F and 160°F, you can make nutrients available more quickly to your garden. This method requires regular turning and monitoring of moisture levels, but it's perfect if you're eager to enrich your soil fast. On the other hand, cold composting is less labor-intensive. It involves simply piling your materials and letting them break down over time, usually taking a year or so. While slower, this method requires minimal effort and still results in nutrient-rich compost.

Delving deeper into composting methods, consider vermicom-posting, an innovative approach where worms play a crucial role in breaking down the manure. This method not only speeds up decomposition but also enhances the nutrient content of the compost. Worms process the manure, leaving behind castings that are rich in essential nutrients and beneficial microbes. These cast-ings can be directly used in your garden, providing a quick and efficient nutrient boost to your plants. Another effective technique is trench composting, which involves burying layers of manure mixed with carbon-rich materials directly in your garden beds. Over time, these layers decompose, gradually enriching the soil with nutrients right where your plants need them. This method is particularly useful for gardeners looking to improve soil fertility without the need for a separate composting area. It's a straightforward, labor-saving approach that mimics natural soil-building processes, offering a practical way to recycle chicken manure directly into the garden.

When it comes to using composted chicken manure in your garden, timing is everything. Apply it during active growth stages when plants can most benefit from the nutrients. Early spring is often ideal for spreading compost before planting your crops. But be careful not to place it directly on plant roots as this can cause damage. Instead, mix it into the top layer of soil or use it as a side dressing around established plants. This approach ensures that nutrients slowly seep into the root zone without overwhelming your plants.

PRACTICAL TIP: TIMING COMPOST APPLICATIONS

To maximize the benefits of your compost, apply it in early spring or fall when the soil is warm and plants are either getting ready to grow or winding down their growth cycle. This timing allows nutrients to integrate into the soil before extreme weather conditions set in.

Using chicken manure as compost in your garden isn't just about recycling waste; it's about creating a sustainable cycle where everything contributes back to the earth. You'll find that with each application, your garden becomes more vibrant and productive. The soil's improved structure and fertility lead to healthier plants and higher yields, whether you're growing tomatoes or sunflowers. It's incredibly satisfying to watch your garden flourish from something that might have otherwise been discarded.

Consider starting small workshops or demonstration plots in your neighborhood to share these enriching practices with others, further nurturing a community committed to sustainability through practical actions.

Incorporating chicken manure into your gardening routine transforms what many see as a nuisance into an asset. With patience and care, you can harness its full potential and contribute to a thriving garden ecosystem. Your plants will thank you with bountiful harvests and robust growth, rewarding every effort you've put into managing this natural resource wisely.

Sharing your journey within local community gardens or social groups can amplify the collective impact, helping to spread awareness of sustainable practices and inviting others to contribute to the ecological well-being of the neighborhood.

REDUCING WASTE THROUGH EFFICIENT CHICKEN-KEEPING

Imagine standing in your backyard, watching your chickens peck contentedly at kitchen scraps tossed their way. This simple act is more than just feeding; it's a step toward reducing waste and promoting sustainability. Kitchen scraps, often destined for the trash, become a valuable resource for your chickens. They relish the variety, and you reduce waste production by incorporating these leftovers into their diet. Not only does this minimize food waste, but it also cuts down on the feed you purchase, saving money and resources. Recycling doesn't stop there; used chicken bedding can be a boon for your garden. Once it's spent in the coop, it transforms into mulch or compost, enriching your soil with nutrients that promote plant growth. This cycle of reuse and recycle illustrates how efficient chicken-keeping can significantly contribute to waste reduction.

When combined with strategies like composting, this feed replacement becomes part of a larger, harmonious cycle, bringing fresh perspectives on sustainable living into your daily routine.

Thinking outside the coop, there are innovative ways to repurpose chicken byproducts like feathers and litter. Feathers can be surprisingly versatile. As mentioned earlier, they can be used in crafts or as insulation material. Their lightweight structure makes them excellent for stuffing pillows or creating decorative items.

Meanwhile, chicken litter, commonly dismissed as mere waste, harbors untapped potential in the realm of biochar production. Chicken litter is the material that lines the floor of your chicken coop and run, and it plays a big role in keeping your flock healthy and your coop clean. At its most basic, chicken litter is a combination of bedding material (like wood shavings, straw, hay, or chopped leaves) and chicken droppings. Over time, this mixture builds up, and how you manage it depends on the system you're using—either regular cleaning or a method like the deep litter method.

The process of biochar production involves pyrolysis, where the litter is subjected to high temperatures in an anaerobic environment, effectively transforming it into biochar. This resultant form of charcoal serves as a powerful soil amendment, significantly enhancing soil fertility when incorporated into your garden beds. The addition of biochar to the soil does more than just dispose of waste efficiently; it also enriches the soil with stable carbon, which has a remarkable capacity to retain nutrients and water. This, in turn, boosts the soil's nutrient-holding capacity, leading to healthier plant growth and more bountiful yields. By integrating chicken litter into biochar production, you not only repurpose a readily available byproduct but also contribute to a cycle of sustainability that benefits your garden's ecosystem.

Conserving resources is a key aspect of sustainable chicken-keeping. Water conservation techniques are particularly crucial. Installing rainwater collection systems can provide a renewable water source for your flock, reducing reliance on municipal supplies. Gravity-fed waterers are another efficient option, allowing chickens to access water as they need it without waste. These systems ensure that every drop counts, especially during dry spells when water conservation becomes even more critical. Energy efficiency in coop design is also vital. Implementing solar panels or using natural light sources minimizes energy consumption while maintaining a comfortable environment for your chickens. Proper insulation reduces the need for artificial heating or cooling, keeping energy use low year-round.

Real-life examples highlight the success of these waste reduction practices. Consider a zero-waste chicken operation that utilizes every aspect of its setup to minimize impact on the environment. By recycling bedding, repurposing feathers, and conserving water and energy, these operations epitomize sustainable chicken-keeping. Testimonials from eco-conscious chicken keepers emphasize the benefits of such practices. One keeper shared how implementing rainwater collection saved hundreds of gallons annually, significantly reducing their environmental footprint. Another described the satisfaction of using biochar in their garden, witnessing improved soil health and plant growth as a direct result.

These stories underscore how reducing waste through efficient chicken-keeping aligns with broader sustainability goals. It's about creating a system where nothing is wasted and everything has a purpose. By embracing these practices, you contribute positively to both your local environment and global efforts to reduce waste and conserve resources. It's a rewarding endeavor that highlights the interconnectedness of chickens, gardens, and sustainable living.

Encourage neighborhood competitions or group activities focused on reducing waste and sharing results, transforming mundane waste-management tasks into fun, community-building initiatives.

These waste reduction strategies not only benefit the environment but also enhance your chicken-keeping experience by fostering a deeper connection with nature and sustainable practices. As you explore ways to minimize waste within your setup, remember that every small change can make a significant impact on both your backyard ecosystem and the planet at large. Embracing these practices means taking steps toward a future where chicken-keeping is not just about raising birds but also about nurturing a sustainable lifestyle that benefits everyone involved.

COMMUNITY ENGAGEMENT AND KNOWLEDGE SHARING

There's something incredibly fulfilling about knowing you're not alone in your chicken-keeping endeavors. Engaging with your local community can transform what might feel like a solitary hobby into a shared passion. Organizing chicken coop tours not only invites neighbors into your world but also opens up dialogue about sustainable practices. It's a chance for others to see firsthand how chickens contribute to a more eco-friendly lifestyle. Hosting workshops on sustainable chicken-keeping practices is another excellent way to foster community spirit. Whether it's teaching others how to build a predator-proof coop or sharing tips on composting chicken manure, these gatherings create bonds and spread knowledge.

Consider branching into local schools or community centers, organizing talks or clubs focused on chicken-keeping and sustainability—sparking curiosity and interest in upcoming generations.

The digital age offers endless opportunities for knowledge exchange. Online platforms like Backyard Chickens serve as vibrant forums where chicken keepers of all levels share advice, troubleshoot issues, and celebrate successes. Joining social media groups dedicated to poultry enthusiasts is like having a virtual coffee chat with friends who share your interests. These spaces are invaluable for finding answers to questions, learning about new trends, and connecting with a global community.

Collaborating on blog posts, joint YouTube videos, or cross-platform projects might open new pathways of knowledge sharing and engagement in your local networks.

Collaboration takes chicken-keeping to the next level. Imagine pooling resources with fellow enthusiasts for co-op purchasing of feed and supplies. Not only does this reduce costs, but it also fosters a sense of camaraderie. Collective composting initiatives are another brilliant idea. By working together, you can manage larger volumes of waste more efficiently, turning potential problems into garden gold.

Success stories abound in communities that embrace shared chicken-keeping efforts. Take, for instance, a neighborhood that transformed a neglected lot into a thriving community garden, complete with chickens. The project not only revitalized the space but also provided fresh produce and eggs for local families. In urban areas, initiatives promoting backyard chickens have gained momentum, proving that even city dwellers can enjoy the benefits of sustainable poultry keeping.

In urban projects, integrate elements like rooftop gardens, vertical plantings, or even tiny coops in balconies, adapting chicken-keeping to the urban context while further encouraging awareness and participation.

This communal approach to chicken-keeping enriches the experience for everyone involved. It creates networks of support, where each participant contributes knowledge and resources for mutual benefit. By engaging in these activities, you become part of a larger movement toward sustainable living. You're not just raising chickens; you're fostering connections and building a community dedicated to environmental stewardship.

Engage with local businesses or sponsors to create sponsorships for events, enhancing collaboration within community projects and bringing a broader awareness to sustainable practices.

CHAPTER CONCLUSION

As we wrap up this chapter on integrating chickens into sustainable living, it's clear that these feathered friends are more than just egg producers—they're catalysts for change. They bring people together, enrich our gardens, and contribute to healthier ecosystems. Whether you're connecting with neighbors or sharing tips online, remember that every step you take toward sustainability makes a difference. In the next chapter, we'll explore the intricacies of advanced chicken care techniques, diving deeper into understanding your flock's needs and behaviors. So, stay tuned as we continue this exciting adventure in chicken-keeping!

CHAPTER 8
ADVANCED TIPS AND TROUBLESHOOTING

HANDLING AGGRESSIVE CHICKENS AND MAINTAINING PEACE

he tranquil morning on a farm paints a picture of utopian peace where the countryside awakens with the gentle rise of the sun. The golden light stretches over fields and coops, promising a day of routine farming joys. Yet, within the confines of the chicken coop, a different drama might be unfolding —a chaos not of nature's making but chickens clashing out of sync with the placidity around. This volatile atmosphere among chickens can be as puzzling as it is unsettling. To restore peace, delving deeper into the causes of such aggression, alongside effective interventions, is paramount.

IDENTIFYING THE ROOTS OF AGGRESSION

Aggression often roots itself in scarcity of space, a common woe that transforms calm into conflict. Each chicken, like humans, harbors an intrinsic need for personal territory. When overcrowded like sardines in a can, their interactions can turn hostile. Drawing parallels to a contest where resources are limited, tension and conflict naturally escalate. In these tight quarters, a simple peck escalates rapidly into repeated brawls. To combat this, enhancing coop spaces or reevaluating flock numbers can alleviate these spatial pressures. Doing so grants each bird the needed breathing room, quelling stress-fueled aggression from the onset.

Furthermore, the disruption of a carefully maintained social hierarchy by introducing new members can lead to turmoil in an otherwise peaceful flock. Chickens, like many social animals, adhere to a pecking order that guides their interactions. Introducing new faces can topple this well-established hierarchy. Acting pre-emptively with measures such as temporary separations or quarantine periods before introducing newcomers can allow your flock to adjust, minimizing these social ripples.

STRATEGIES FOR CONFLICT RESOLUTION

Forging peace among chickens requires a strategic approach, blending gentle handling with thoughtful strategy. Temporary isolation stands as a potent method for defusing tensions among overly aggressive individuals. Remove the belligerents and allow the flock to settle into a rhythm that meshes peace over chaos. Reintroduce them gradually, ideally alongside treats or calm periods of interaction, to ensure smoother transitions and acceptance back into the pecking order without innate hostility.

Creating an engaging environment within the coop can also displace aggression. Stimulating chickens' curiosity sways their focus from conflict to exploration. Constructing a coop with varying levels and the inclusion of diverse structural elements can address boredom and aggression alike. Providing dust baths offers chickens not only a hygienic ritual but also a distraction, while perches facilitate natural behaviors, redirecting potential aggression.

Implementing behavioral modification tools and techniques

could act as a supplementary aid for maintaining peace. The use of peepers or blindfolds to obscure a chicken's direct line of sight proves beneficial in reducing pecking tendencies, albeit appearing unconventional at first. These devices ensure the bird is not harmed, but rather aids in curbing aggressive behavior. This approach can be reinforced with positive reinforcement, rewarding peaceful behavior with treats, thereby nudging their default response towards calmness.

SUCCESS STORIES AND EXPERT INSIGHTS

Instances abound highlighting the transformations feasible with these methodologies. Consider Sarah, a diligent chicken keeper who battled ceaseless turmoil within her flock following the integration of new hens. Her toolkit included temporary separations, clever distractions like mirrors, and homemade hanging toys. These measures gradually dissolved tensions, fostering an environment of calmness. Key to her success was persistence and insightful observation, allowing her to tweak interventions as needed tailored to her unique environment.

Experts in the field echo the importance of deciphering the individual personalities present within a flock. Each bird possesses distinctive traits and quirks, urging a nuanced approach to their care. For instance, more assertive breeds might require stricter boundaries than their gentle counterparts. Maintaining a behavioral log for the flock can provide insights into patterns that necessitate intervention, thus cultivating a harmonious environment for all.

CASE STUDY: RESTORING HARMONY IN A TROUBLED FLOCK

Take Tom, who was confronted with an unruly rooster that disrupted the equilibrium of his flock. Through observation, Tom honed his skills in decoding behavioral cues, deciding to isolate the rooster temporarily while enhancing the coop environment with enriching activities. Dust baths and hanging treats transformed the once disharmonious setting into one of serenity.

Upon cautiously reintroducing the rooster to the flock, Tom took meticulous steps to ensure a smooth transition. He observed the interactions, ready to intervene at the first sign of aggression. Utilizing positive reinforcement techniques, such as offering treats during moments of calm behavior and gently guiding the rooster towards less dominant hens, Tom fostered a nurturing environment. This strategic approach enabled the rooster and the hens to gradually adjust to each other's presence, easing tensions and allowing the flock to organically reestablish its natural hierarchy and social structure.

By embedding these strategies within everyday practice, you nurture an environment where peace flourishes, to the benefit of both chickens and keeper. Each flock encapsulates its unique dynamics, fostering a need for keen observation and adaptability. Exercising patience, creativity, and empathy holds the key to overcoming challenges on your chicken-keeping adventure. Reflecting on the ongoing experiences and fine-tuning techniques as circumstances shift offer pathways to genuine harmony amid feathers and clucks.

CREATIVE SOLUTIONS FOR COMMON CHICKEN-KEEPING CHALLENGES

Venturing into your coop only to discover chickens pecking on their eggs as an impromptu brunch buffet is undoubtedly a frustrating scenario. Fortunately, inventive solutions abound for this conundrum. A fundamental yet effective measure is frequent egg collection throughout the day, which diminishes the opportunity for eggs to become targets of curiosity or hunger. Alternatively, installing roll-away nesting boxes ensures eggs safely roll into a protected compartment upon being laid, thereby keeping them out of chickens' reach.

Feather pecking presents another persistent challenge that transcends superficial annoyances to potentially cause stress and injury. Anti-pecking sprays, with their bitter taste, discourage such behavior by making the act undesirable. In severe cases, identifying the instigator allows for targeted interventions, damping the coop's stress levels. Engaging toys or mirrors furnish mental stimulation, aiding in alleviating boredom and reducing the urge to indulge in feather-pecking.

Noise, often a byproduct of chicken coops, may rile neighbors in proximity. Planting hedges around your yard creates a natural noise barrier, muting the clucks and crows that amplify in open spaces. To further contain sound, consider incorporating acoustic elements within the coop, such as insulated panels or acoustic foam. These materials can absorb sound without interfering with necessary ventilation, harmonizing the coop's acoustic landscape.

Maintain impeccable hygiene to uphold flock health; it needn't become a cumbersome task. Utilizing UV lamps can keep bacteria at bay on surfaces, serving as a chemical-free answer to sanitation. Scheduling deep-cleans that involve thorough bedding removal and surface disinfection prevents pathogen buildup. Eco-friendly disinfectants can be harnessed to advance hygiene goals with minimal adverse environmental effects.

INNOVATIVE COOP ENHANCEMENTS FOR IMPROVED FUNCTIONALITY

The farming realm is not left behind in the march of technological advancement. Automated feeder and watering systems afford ease by ensuring chickens' needs are met in your absence. These innovations can be tailored through timers or responsiveness, guaranteeing ready access to sustenance at all times. Moreover, smart lighting and temperature control systems introduce automated management of environmental factors, preserving egg production consistency and flock well-being.

Modular design elements ring in flexibility for chicken coops, featuring removable panels that allow straightforward access for maintenance. Expandable segments accommodate your flock's growth, thus assuring adequate space and comfort as bird numbers escalate. This adaptive framework ensures your coop's functionality persists despite changes in flock size while merging style with practical needs.

Incorporating sustainability in coop functionalities is increasingly paramount. Solar panels provide energy independence by powering lights or warming elements. Rainwater harvesting systems offer an eco-friendly method to supply water, easing dependency on municipal sources. Switching to renewable resources aligns practices with sustainability, enhancing the longevity and independence of your coop systems.

In the bustling environments of urban and suburban areas, where space is at a premium, innovative chicken coop designs have risen to the challenge. Vertical configurations extend the living spaces of chickens upwards, maximizing the limited ground area available. These coops are not only practical but also cleverly incorporate storage spaces for feed, tools, and supplies, ensuring everything needed for chicken care is neatly organized and easily accessible. Such architectural innovations seamlessly integrate into the smaller spaces of city living, offering a perfect blend of functionality and visual appeal.

Implementing these enhancements enriches not only the coop's functionality but also the chicken-keeping experience itself. The integration of technology, modular adaptability, sustainability, and creativity fosters a coop environment supporting both avian and human needs harmoniously. Whether rooted in countryside expanses or city confines, these creative approaches translate into a more efficient setup. The result is a flourishing haven—where hens lay and cluck serenely, offering the simple pleasures of fresh eggs and companionship secured in tranquility.

KEEP THE FLOCK GROWING

Now that you have everything you need to raise happy, healthy chickens, it's time to pass on what you've learned and help others get started, too.

By leaving your honest review of this book on Amazon, you'll be showing other new chicken-keepers where they can find the guidance they need—and helping them feel confident from day one.

Thank you for being part of this growing backyard chicken community.

Raising chickens stays strong when we share what we know— and you're helping me do just that.

👉 **Scan the QR code or go here to leave your review on Amazon:** https://www.amazon.com/review/review-your-purchases/?asin=B0F34TZ1RK

Thanks again for being here. I'm cheering you on—every step of the way!

— Avery Sage

CONCLUSION

As you reach the end of this guide, take a moment to reflect on the journey we've shared. From those first tentative steps into the world of chicken-keeping to mastering the nuances of flock management, you've come a long way. We began with the basics—choosing the right breeds, setting up a coop, and understanding local regulations. As you progressed, so did your knowledge and confidence. You delved into the intricacies of chicken behavior, health management, and even advanced techniques to create a thriving, sustainable ecosystem in your own backyard.

Throughout this book, my vision has been to empower you, a beginner, to not only raise chickens but to do so with joy and success. I wanted to equip you with the skills to gather fresh eggs and relish the satisfaction of sustainable living, all while connecting with nature. I hope you now feel equipped to achieve just that.

Let's revisit some key takeaways from our shared journey. You've learned how to recognize and respond to chicken behaviors, ensuring a harmonious flock. You've gained insights into managing health and wellness, from vaccinating to recognizing signs of stress and disease. You've discovered how to integrate chickens into a permaculture system, enriching both your garden and your flock's lives. Each chapter was designed to build your confidence and provide actionable steps to navigate any challenges you might face.

The benefits of keeping chickens are plentiful, and I hope you've felt the excitement of fresh eggs, straight from your coop. Beyond the eggs, though, you've embraced sustainable practices, reducing waste and enhancing your garden's fertility. And let's not overlook the personal satisfaction—the simple, profound joy of watching your chickens thrive, knowing you've played a part in their well-being.

As you move forward, carry with you the confidence that you can manage a healthy, productive flock. Remember, challenges will arise, but with the knowledge and tools you've gained here, you're well-prepared to tackle them head-on. Trust in your abilities, and know that mistakes are merely opportunities for learning and growth.

Now, it's time to take action. Whether you're just starting out or looking to enhance your current setup, use this book as your trusted guide. Begin by taking small steps, and gradually build on your achievements. Each step you take brings you closer to a flourishing flock and a more sustainable lifestyle.

I encourage you to engage with the broader chicken-keeping community. Join local groups or online forums to share your experiences and learn from fellow enthusiasts. These connections will enrich your journey, offering support and camaraderie as you continue to explore the world of chicken-keeping.

Reflect on your personal growth and accomplishments. Consider how far you've come since the beginning of this book. Each experience, each challenge overcome, has contributed to your development as a chicken keeper. Celebrate these milestones, and let them motivate you to keep growing.

Finally, I want to express my deepest gratitude. Thank you for placing your trust in this guide and for your commitment to this rewarding journey. It's been a privilege to accompany you, and I look forward to hearing about your successes. Here's to many happy mornings spent gathering fresh eggs and enjoying the delightful company of your chickens.

REFERENCES

1. camrynrabideau.com. (2024). *The Best Chicken Breeds For Beginners - Camryn Rabideau*. https://camrynrabideau.com/2024/03/27/the-best-chicken-breeds-for-beginners/

2. poultry.extension.org. (n.d.). *DEVELOPING REGULATIONS FOR KEEPING URBAN*. https://poultry.extension.org/articles/poultry-management/urban-poultry/developing-regulations-for-keeping-urban-chickens/

3. strombergschickens.com. (n.d.). *Essential Equipment For a Chicken Coop*. https://www.strombergschickens.com/blog/essential-equipment-for-a-chicken-coop/?srsltid=AfmBOopZEhrnf3wpGWdp5e YC0FT0VVuYN9OzB7ywinJub6s7BG6l6LVf

4. thespruce.com. (n.d.). *25 Free Chicken Coop Plans*. https://www.thespruce.com/free-chicken-coop-plans-1357113

5. lifeatcobblehillfarm.com. (2023). *Tips For Predator Proofing Your Chicken Coop*. https://www.lifeatcobblehillfarm.com/2023/09/tips-for-predator-proofing-your-chicken.html

6. blog.meyerhatchery.com. (2019). *Coop Ventilation and Why it is Important | Meyer Hatchery Blog*. https://blog.meyerhatchery.com/2019/12/ventilation-in-your-coop-and-why-it-is-important/#:~:text=Vents%20placed%20high%20above%20your,prevent%20additional%20moisture%20build%2Dup.

7. somerzby.com.au. (n.d.). *Chicken Coop Maintenance - A Step-by-Step Guide*. https://www.somerzby.com.au/blog/chicken-coop-maintenance/?srsltid=AfmBOor6BjTdVebea5gIlDhQDheMYgo-SBqOe71KY8EkhBQj6CweTTl_

8. merckvetmanual.com. (n.d.). *Nutritional Requirements of Poultry - Merck Veterinary Manual*. https://www.merckvetmanual.com/poultry/nutrition-and-management-poultry/nutritional-requirements-of-poultry

9. farmstandapp.com. (n.d.). *10 Best Organic Feeds for Raising Chickens That Promote ....* https://www.farmstandapp.com/6983/best-organic-feeds-for-raising-chickens/

10. cacklehatchery.com. (n.d.). *Safe Kitchen Scraps for Chickens - Cackle Hatchery*. https://www.cacklehatchery.com/safe-kitchen-scraps-for-

chickens/#:~:text=Soft%20items%20like%20cucumbers%20and,safe%20kitchen%20scraps%20for%20chickens.

11. flygrubs.com. (n.d.). *Seasonal Chicken Feeding Guide for Summer and Winter* .... https://flygrubs.com/blogs/news/seasonal-chicken-feeding-guide

12. merckvetmanual.com. (n.d.). *Common Infectious Diseases in Backyard Poultry*. https://www.merckvetmanual.com/exotic-and-laboratory-animals/backyard-poultry/common-infectious-diseases-in-backyard-poultry

13. backyardpoultry.iamcountryside.com. (n.d.). *3 Herbs to Heal and Prevent Chicken Respiratory Infections.* https://backyardpoultry.iamcountryside.com/feed-health/3-herbs-to-heal-and-prevent-chicken-respiratory-infections/

14. extension.psu.edu. (n.d.). *Deworming Backyard Poultry.* https://extension.psu.edu/deworming-backyard-poultry

15. merckvetmanual.com. (n.d.). *Vaccination of Backyard Poultry - Exotic and Laboratory Animals.* https://www.merckvetmanual.com/exotic-and-laboratory-animals/backyard-poultry/vaccination-of-backyard-poultry

16. ncbi.nlm.nih.gov. (n.d.). *Poultry - Effect of Environment on Nutrient Requirements* https://www.ncbi.nlm.nih.gov/books/NBK232332/

17. khpet.com. (n.d.). *10 of the Best Chicken Breeds for Eggs.* https://khpet.com/blogs/farm/10-of-the-best-chicken-breeds-for-eggs?srsltid=AfmBOopERpQYx4s7DmV_8PVnPWIGs3FOuy5NJj6ZM3TI4GF7HNTes3re

18. lalasfarm.com. (n.d.). *Chicken Molting: Nutritional Considerations, Potential* .... https://www.lalasfarm.com/post/chicken-molting-nutritional-considerations-potential-discomfort-and-insights-into-the-molt-season

19. farmhouseonboone.com. (n.d.). *Water Glassing Eggs - Farmhouse on Boone.* https://www.farmhouseonboone.com/water-glassing-eggs/#:~:text=Water%20glassing%20eggs%20involves%20submerging,day%20the%20hen%20laid%20them.

20. grubblyfarms.com. (n.d.). *5 Steps to Winterize Your Chicken Coop.* https://grubblyfarms.com/blogs/the-flyer/winterize-your-chicken-coop?srsltid=AfmBOoqMq1j-8EXHHRZyUCKAYJFGmIrKKWTFSPYZccaUKKVuzAXgK8QH

21. backyardpoultry.iamcountryside.com. (n.d.). *What is the Best Feed for Chickens in Summer?.* https://backyardpoultry.iamcountryside.com/feed-health/what-is-the-best-feed-for-chickens-in-summer/

22. the-chicken-chick.com. (n.d.). *11+ Tips for Predator-proofing Chickens*. https://the-chicken-chick.com/11-tips-for-predator-proofing-chickens/

23. chickenwhisperermagazine.com. (n.d.). *Keeping Chickens Comfy During Seasonal Changes*. https://chickenwhisperermagazine.com/the-chicken-movement/keeping-chickens-comfy-during-seasonal-changes/#:~:text=Sudden%20temperature%20changes%20can%20affect,signs%20indicate%20stress%20and%20illness.

24. strombergschickens.com. (n.d.). *Guide to Free-Range Farming: Facts, Pros, and Cons*. https://www.strombergschickens.com/guide-to-free-range-farming-facts-pros-and-cons/?srsltid=AfmBOoo61KfwMveU1wo_BlYitnBDAHqo3cbs_DLpFCpdYU5nnC_oUjha

25. thecritterdepot.com. (n.d.). *Integrating Chickens into Permaculture Gardens: A Guide ....* https://www.thecritterdepot.com/blogs/news/integrating-chickens-into-permaculture-gardens-a-guide-for-sustainable-living

26. extension.unr.edu. (n.d.). *Using Chicken Manure Safely in Home Gardens and ....* https://extension.unr.edu/publication.aspx?PubID=3028#:~:text=It%20should%20be%20composted%20or,around%20plants%2C%20people%20and%20pets.

27. econourish.co.uk. (n.d.). *10 Eco Tips for More Sustainable Chicken Keeping*. https://econourish.co.uk/sustainable-chicken-keeping-10-tips/

28. hobbyfarms.com. (n.d.). *How To Deal With Aggressive Chicken Behavior*. https://www.hobbyfarms.com/aggressive-chickens-behavior-tips/

29. backyardchickens.com. (n.d.). *Chicken Coops*. https://www.backyardchickens.com/articles/categories/chicken-coops.12/

30. gardenculturemagazine.com. (n.d.). *Permaculture Diaries: Why Raising Chickens in the Forest ....* https://gardenculturemagazine.com/permaculture-diaries-why-raising-chickens-in-the-forest-is-best/

31. hobbyfarms.com. (n.d.). *These Apps Can Help You Manage Backyard Chickens*. https://www.hobbyfarms.com/these-apps-can-help-you-manage-backyard-chickens/

www.ingramcontent.com/pod-product-compliance
Lightning Source LLC
Chambersburg PA
CBHW031859200326
41597CB00012B/483